T0137074

Sustainable Civil Infrastructures

Sustainable Infrastructure impacts our well-being and day-to-day lives. The infrastructures we are building today will shape our lives tomorrow. The complex and diverse nature of the impacts due to weather extremes on transportation and civil infrastructures can be seen in our roadways, bridges, and buildings. Extreme summer temperatures, droughts, flash floods, and rising numbers of freeze-thaw cycles pose challenges for civil infrastructure and can endanger public safety. We constantly hear how civil infrastructures need constant attention, preservation, and upgrading. Such improvements and developments would obviously benefit from our desired book series that provide sustainable engineering materials and designs. The economic impact is huge and much research has been conducted worldwide. The future holds many opportunities, not only for researchers in a given country, but also for the worldwide field engineers who apply and implement these technologies. We believe that no approach can succeed if it does not unite the efforts of various engineering disciplines from all over the world under one umbrella to offer a beacon of modern solutions to the global infrastructure. Experts from the various engineering disciplines around the globe will participate in this series, including: Geotechnical, Geological, Geoscience, Petroleum, Structural, Transportation, Bridge, Infrastructure, Energy, Architectural, Chemical and Materials, and other related Engineering disciplines.

More information about this series at http://www.springer.com/series/15140

Sayed Hemeda · Mounir Bouassida
Editors

Contemporary Issues in Soil Mechanics

Proceedings of the 2nd GeoMEast International Congress and Exhibition on Sustainable Civil Infrastructures, Egypt 2018 – The Official International Congress of the Soil-Structure Interaction Group in Egypt (SSIGE)

Editors
Sayed Hemeda
Cairo University
Cairo, Egypt

Mounir Bouassida
University of Tunis El Manar
Tunis, Tunisia

ISSN 2366-3405 ISSN 2366-3413 (electronic)
Sustainable Civil Infrastructures
ISBN 978-3-030-01940-2 ISBN 978-3-030-01941-9 (eBook)
https://doi.org/10.1007/978-3-030-01941-9

Library of Congress Control Number: 2018957281

This Springer imprint is published by the registered company Springer Nature Switzerland AG
The registered company address is: Gewerbestrasse 11, 6330 Cham, Switzerland

Contents

About the Editors

M. Bouassida is a professor of civil engineering at the "Ecole Nationale d'Ingénieurs de Tunis" of the "Université de Tunis El Manar" where he earned his B.S., M.S., Ph.D., and doctorate of sciences diplomas, all in civil engineering. He is the director of the Research Laboratory in Geotechnical Engineering and has supervised 16 Ph.D. and 29 master of science graduates. His research focuses on soil improvement techniques and behaviour of soft clays. He is the (co) author of 87 papers in refereed international journals; 126 papers, including 18 keynote lectures; and three books. He is a member of the editorial committees of journals Ground Improvement (ICE), Geotechnical Geological Engineering, Infrastructure Innovative Solutions, and International Journal of Geomechanics (ASCE). He is also an active reviewer in several international journals. As a 2006 Fulbright scholar, he created a novel methodology for the design of foundations on reinforced soil by columns. He was awarded the 2006 S. Prakash Prize for Excellence in the practice of geotechnical engineering. In 2008, he launched a Tunisian consulting office in geotechnical engineering, SIMPRO. He is a co-developer of the software Columns 1.01 used for designing column-reinforced foundations. He held the office of the vice-president of ISSMGE for Africa (2005–2009). He benefited from several grants as a visiting professor in the USA, France, Belgium, Australia, Vietnam, Hong Kong, and Norway.

Dr. Sayed Hemeda is a Doctor of Civil Engineering Department, Aristotle University of Thessaloniki, Greece. He occupies the position of Associate Professor, Conservation Department, Faculty of Archaeology, and Cairo University, Egypt. Also, he is the vice manager of conservation centre. He was awarded the Cairo University's prize of scientific excellence in 2017. He was awarded the Cairo University's prize of encourage in 2014. He was awarded the Cairo University's prize for the best Ph.D. theses in Faculty of Archaeology, 2009–2010. He published more than 48 international scientific articles in international scientific journals in Springer and Elsevier. He composed 12 scientific international books published in London, Croatia, Roma, and Berlin. He is Editorial Board Member of the Sustainable Civil Infrastructures, book series, published by SpringerNature. He is Editorial Board Member of the Progress of Electrical and Electronic Engineering in Singapore. He is Editorial Board Member of the Geoscience Journal in Singapore. He is Editorial Board Member of the Alexandria Engineering Journal.

Comparison of Prediction Models for the Permeability of Granular Materials Using a Database

Shuyin Feng(✉), Paul J. Vardanega, and Erdin Ibraim

Department of Civil Engineering, University of Bristol, Bristol, UK
shuyin.feng@bristol.ac.uk

Abstract. The hydraulic conductivity characteristics of the materials which comprise pavement structures are linked to in service performance. This paper briefly reviews a series of well-known models to predict hydraulic conductivity. An approach which makes use of the grading entropy coordinates is also studied. The database includes information on the gradation, hydraulic conductivity and porosity characteristics for over 150 gravel mixtures. Comparison of the studied models reveals that the 'Kozeny-Carman' model gives the best predictions when considering the entire database. The results of the regression analysis reveal that for granular mixtures comprising greater than 50% sand, the 'Shepherd' or 'Hazen' approaches may be preferred. However, for mixtures comprising less than with 50% sand, the 'Kozeny-Carman' and 'grading entropy' approaches are preferred.

1 Introduction

Water ingress into road pavements is potentially detrimental to the mechanical properties of the structure (e.g., Kandhal and Rickards 2001; Mallick and El-Korchi 2008; Thom 2014; Ghabchi et al. 2015). Engineers need to make rapid assessments of the hydraulic conductivity of pavement materials. This paper follows a detailed laboratory study which compared the 'Hazen', 'Shepherd', 'Kozeny-Carman' and 'Chapuis' models along with a regression model that uses the grading entropy coordinates (Feng 2017; Feng et al. 2018a). Feng (2017) and Feng et al. (2018a) presented results on only one material but testing was conducted over a wide range gradations. This paper aims to investigate the relative merits for the aforementioned models using a larger database of experimental observations.

1.1 Traditional Models for Hydraulic Conductivity

The hydraulic conductivity k (in Length.Time^{-1}) is given by (e.g., Craig 2004, p. 31):

$$k = \frac{\gamma}{\mu} K \tag{1}$$

S. Hemeda and M. Bouassida (Eds.): GeoMEast 2018, SUCI, pp. 1–13, 2019.
https://doi.org/10.1007/978-3-030-01941-9_1

Where μ (in Mass.Time^{-1}.Length^{-1}) are the unit weight and the dynamic viscosity of the permeant respectively, and K (in Length^2) is the intrinsic permeability.

The 'Hazen' formula (e.g., Hazen 1893, p. 553) is one of the most commonly-used models which can be expressed as follows:

$$k = (0.7 + 0.03t)C_H D_{10}^2 \quad (2)$$

where t is the temperature in degrees Celsius, D_{10} represents the sieve aperture through which ten percent of the material passes, and C_H is an empirical coefficient.

Shepherd (1989) presented a modification to the 'Hazen' approach using an equation of the following form:

$$k = C_{HS} d_{eff}^a \quad (3)$$

where a (for most materials) generally varies from 1.65 to 1.85 (according to Shepheard 1989), C_{HS} is a regression constant, d_{eff} is the representative particle size (which for consistency is taken in this paper to be D_{10}).

The 'Kozeny-Carman' formulation (Carman 1937, 1939; Kozeny 1927) is another well-known semi-empirical model to describe hydraulic conductivity. Carrier (2003) gives a variant of this formulation, shown as Eq. (4):

$$k = \left(\frac{\gamma}{\mu}\right)\left(\frac{1}{C_{K-C}}\right)\left(\frac{1}{S_A^2}\right)\left(\frac{e^3}{1+e}\right) \quad (4)$$

where $C_{K\text{-}C}$ is an empirical coefficient (which can be taken as approximately 5), S_A is the specific surface area per unit volume of particles, and e is the void ratio.

Chapuis (2004, 2012) developed an amalgamated permeability model for non-plastic sand by combining aspects of the 'Kozeny-Carman' and 'Hazen' approaches. Using a database, Chapuis (2004) found Eq. (5) statistically:

$$k(cm/s) = 2.4622[D_{10}^2 e^3/(1+e)]^{0.7825} \quad (5)$$

where the D_{10} is in mm. Chapuis (2004) suggested that Eq. (5) only provides good predictions for natural soils with 0.003 mm $< D_{10} <$ 3 mm and $0.3 < e < 1$ (137 out of 166 of the collected database in this study fit both criteria).

1.2 The Grading Entropy Method

The 'grading entropy' approach allows a grading curve to be represented vectorially on the normalised grading entropy diagram (Fig. 1). The coordinates are calculated using:

$$A = \frac{\sum_{i=1}^{N} x_i(i-1)}{N-1}. \quad (6)$$

$$B = -\frac{\sum_{i=1}^{N} x_i \log_2 x_i}{\log N}. \quad (7)$$

where '*A*' is the relative base entropy, '*B*' is the normalised entropy increment, *N* is the fraction number, and x_i is the relative frequency of fraction *i*. Lőrincz et al. (2005) and Singh (2014) provide more extensive commentary on the origins and details of the 'grading entropy' approach. Using a similar database to that presented in Vardanega et al. (2017), Feng et al. (2018b) performed multiple linear regression analysis to a large database of hydraulic conductivity measurements on asphalt concrete using the grading entropy co-ordinates as predictors. Feng et al. (2018a) performed similar analysis for a set of laboratory data on a single gravel and found Eq. (8):

$$k20\,^{\circ}\mathrm{C}\,(\mathrm{mm/s}) = 145.47A^{8.90}B^{-2.30} \quad r = 0.95, R^2 = 0.90, n = 30, p < 0.0001 \tag{8}$$

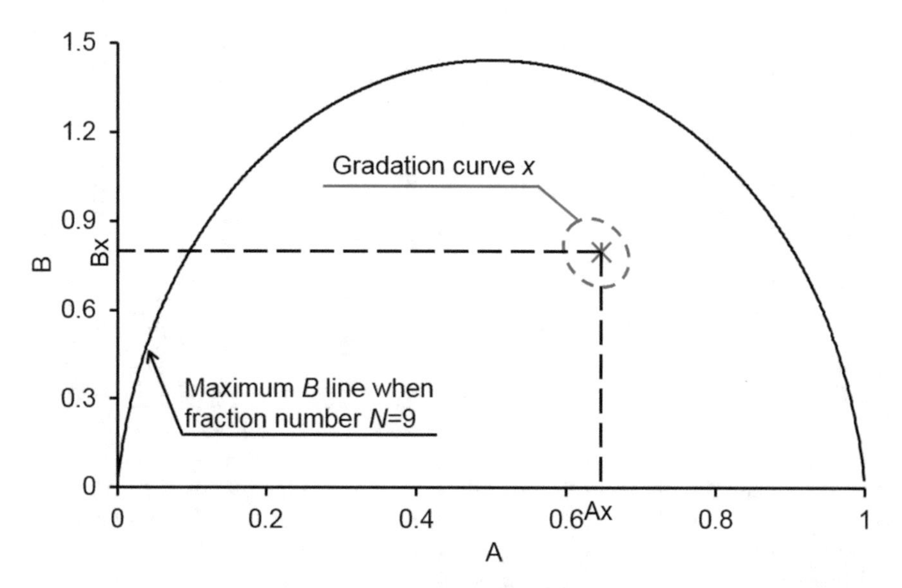

Fig. 1. Sketch example of gradation curve interpretation on normalised grading entropy diagram

Like the 'Hazen' and 'Shepherd' approaches no measurement of void ratio is included in this approach.

2 Database

Table 1 shows the list of ten publications used to source the information for inclusion in the database studied in this paper. 164 hydraulic conductivity tests are included in the database. The test methods (where stated in the original publication), material types, void ratios and gradation parameters of each data source are given in Table 1. The collected database comprises mostly sands and gravels with D_{10} generally ranging from 0.001 mm to 10 mm and void ratio (*e*) ranging from 0.23 to 1.13. The collected hydraulic conductivity data were converted to intrinsic permeability *K* (in mm^2) based on the water temperatures reported in the original publications (ranging from 10 to 20 °C).

Table 1. Summary of database

No.	Sources	Air voids testing method	Hydraulic conductivity testing method	Materials type as described in the original publication	D_{10} (mm)	C_{-U} (mm)	e	n
1	Cabalar and Akbulut (2016)	–	Constant head test	Narli sand	0.10–2.29	1.20–4.22	0.52–0.87	16
				Crushed stone sand	0.10–2.29	1.20–4.22	0.67–1.02	16
2	Indraratna et al. (2012)	–	Constant head test (ASTM D2434-68)	River sand	0.3	1.51–4.03	0.61–0.71	6
3	Mavis and Wilsey (1936)	Geometric	Constant head test	Iowa river sand	0.22–1.81	1.73–5.54	0.50–0.73	12
4	Morris and Johnson (1967)	Geometric	Constant/variable head test	Water laid gravel	1.37–5.45	1.70–2.36	0.61–0.79	3
				Water laid sand	0.001–0.69	1.56–54.39	0.39–0.81	4
5	Wang et al. (2017)	–	Constant head penetration test	Calcareous soil	0.03–0.14	3.30–10.00	0.73–1.13	20
6	Xiao et al. (2013)	–	Gas permeameter test	Crushed aggregate	0.06–1.81	5.67–142.33	0.29–0.35	5
7	Yin et al. (1998)	–	Constant head test	Single-sized crushed stones	2.35–11.00	1.42–1.92	0.40–0.44	3
				Sand	0.15–0.17	2.00–4.76	0.45–0.57	2
				Mechanical stabilized crushed stones	0.14–0.23	15.86–36.51	0.23–0.26	4
8	Dolzyk and Chmielewska (2014)	–	Constant head test	Soil	0.07–0.34	1.95–60.80	0.3–0.923	24
9	Goetz (1971)	–	Constant head test (ASTM D2434-68)	20–30 ottawa sand	0.61	1.20	0.56–0.72	8
				2 ns concrete sand	0.22	4.62	0.37–0.51	4
				Dune sand	0.13	1.88	0.61–0.77	3
				22a gravel	0.15	31.17	0.25–0.41	4
10	Feng (2017) and Feng et al. (2018a)	WA732.2-2011/BS1377:2-1990	Constant head test (BS 1377-5:1990)	Road construction material (mostly crushed aggregate)	0.72–7.02	1.51–7.29	0.53–0.85	30
	Total							164

3 Analysis

3.1 Complete Database

Data from different sources are shown with different markers on Figs. 2, 3, 4, 5 and 6. The numbering of the data entries follows that given in Table 1.

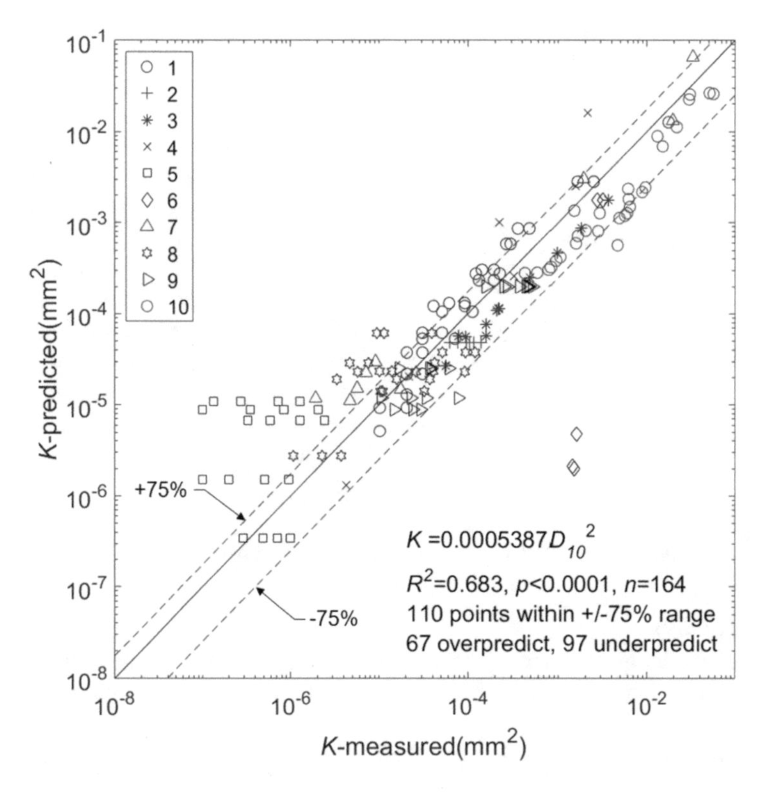

Fig. 2. *K*-predicted versus *K*-measured using Hazen's formula

Hazen (1893)
Figure 2 shows the regression function linking between *K* and D_{10}^2 for the database:

$$K(\text{mm}^2) = 0.00054 D_{10}^2 r = 0.82,\ R^2 = 0.68, n = 164,\ p < 0.0001 \quad (9)$$

The *K*-predicted is plotted against *K*-measured in Fig. 2. The plot shows that 110 out of 164 data points lie between the ±75% bounds (red dashed lines), about 40.8% of the total data points fall above the line of equality (45 degree line) while around 59.2% of the points fall beneath. Figure 2 indicates Eq. (9) generally under-predicts *K*.

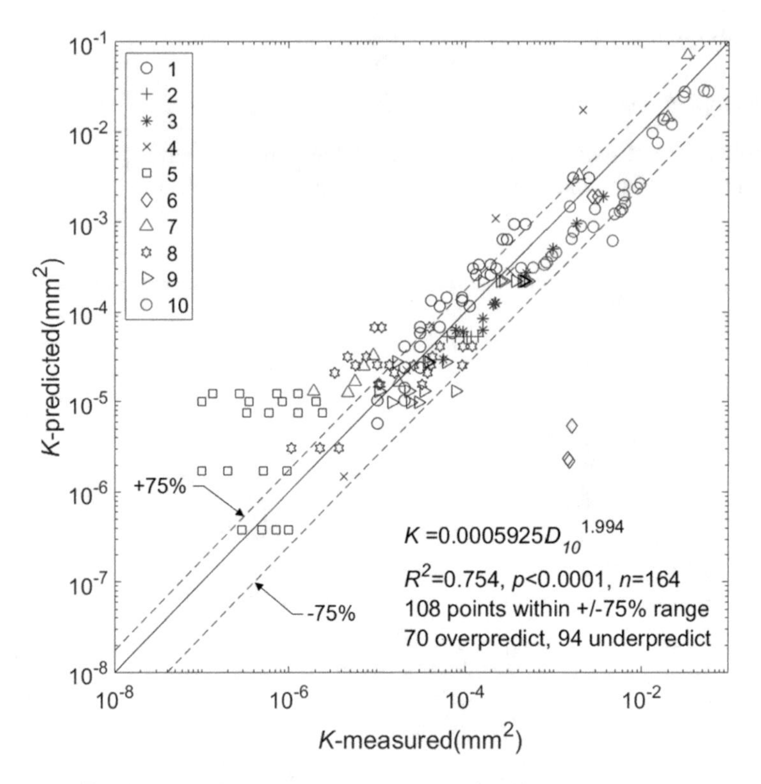

Fig. 3. *K*-predicted versus *K*-measured using Shepherd's model

Shepherd (1989)
'Shepherd' model gives essentially the same result as Eq. (9) (for the data analysed in this paper), the regression result between lnK and lnD_{10} can be rearranged to:

$$K(\text{mm}^2) = 0.00059D_{10}^{1.99} r = 0.87, R^2 = 0.75, n = 164, p < 0.0001 \tag{10}$$

The *K*-predicted versus *K*-measured (Fig. 3) shows that 108 out of 164 of the total points fall within the ±75% margin, 42.7% of the total points lie above the line of equality, while the remainder (57.3%) of the points lie below. Similar to Eq. (9), the Eq. (10) also slightly under-predicts the value in this database.

Kozeny-Carman (1937)
Regression of K with $\frac{\gamma}{\mu} \cdot \frac{1}{S_A^2} \cdot \frac{e^3}{1+e}$ (denoted as *K-C* function in Fig. 3) gave Eq. (11):

$$K(\text{mm}^2) = 0.015 \frac{\gamma}{\mu} \times \frac{1}{S_A^2} \times \frac{e^3}{1+e} r = 0.97, R^2 = 0.93, n = 164, p < 0.0001 \tag{11}$$

where the specific surface (S_A) is calculated based on the method introduced by Chapuis and Légaré (1992). The corresponding predicted versus measured plot (Fig. 4) shows that 97 out of 164 of the total points locate within the ±75% margins, 61.0% of the data

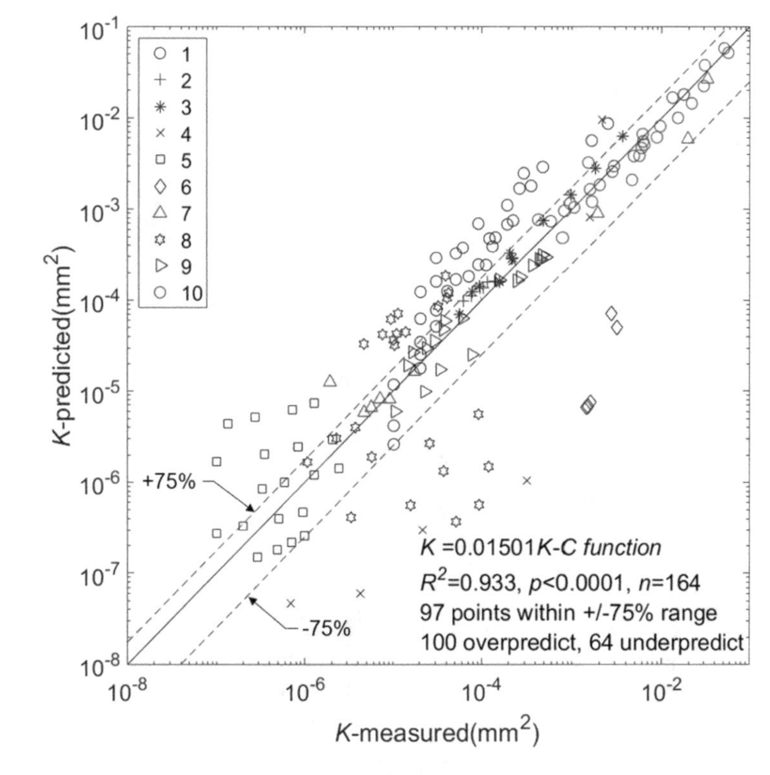

Fig. 4. *K*-predicted versus *K*-measured using Kozeny-Carman's function

points lie above the line of equality while the rest of the data points fall beneath. Equation (11) generally over-predicts the measurements contained in this database.

Chapuis (2004)
The fitted correlation linking lnK and $ln[D_{10}^2e^3/(1+e)]$ can be rearranged to give:

$$K(\mathrm{mm}^2) = 0.0023\left(\frac{D_{10}^2 e^3}{1+e}\right)^{0.85} \quad r = 0.79, R^2 = 0.63, n = 164, p < 0.0001 \quad (12)$$

Equation (12) can be converted using the units adopted in Chapuis (2004, 2012):

$$k(\mathrm{cm/s}) = 2.254[D_{10}^2 e^3/(1+e)]^{0.85} \quad (13)$$

The coefficient of 2.254 and the exponent of 0.85 in Eq. (13) generally conform with the finding of Chapuis (2004, 2012) who gives 2.4622 and 0.7825 for the coefficient and exponent (Eq. 5), note only 1 data source is common between two databases analysed. The predicted versus measured plot (Fig. 5) shows that 78 out of 164 of the total data points fall between the ±75% margins and 43.9% of the data points lie above the line of quality while the remainder of the data points (56.1%) fall beneath, which indicates that the Eq. (12) gives a slightly skewed (under-predicted) prediction of *K*.

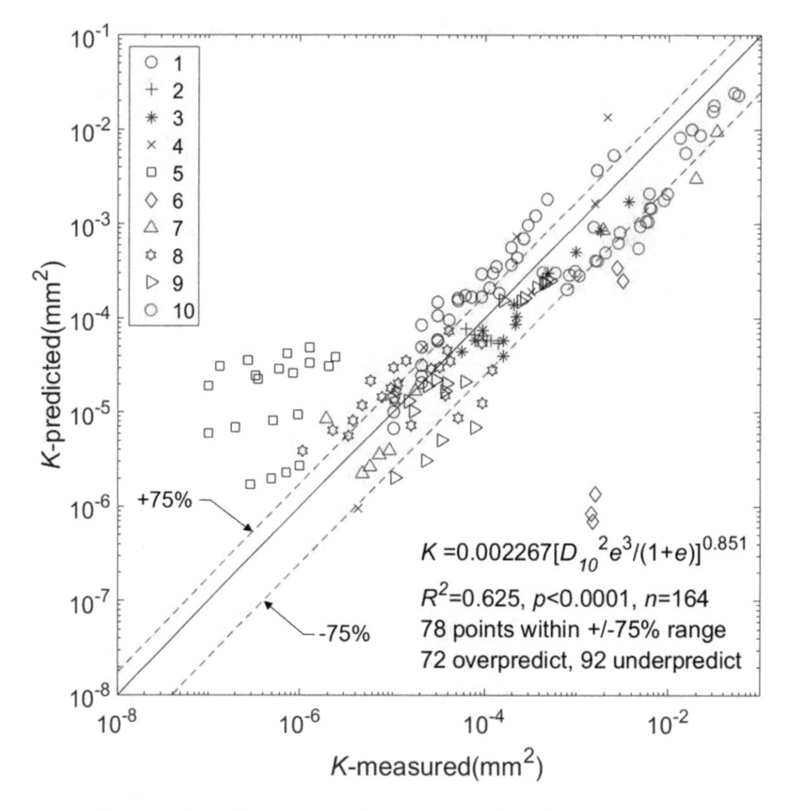

Fig. 5. *K*-predicted versus *K*-measured using Chapuis's model

Grading Entropy Model

The normalized grading entropy coordinates *A* and *B* of the whole database were calculated using Eqs. (6), and (7). The calculated *A* and *B* are then regressed against *K*, the result gives:

$$K(\text{mm}^2) = 0.0048A^{6.04}B^{-0.34}r = 0.83, R^2 = 0.69, n = 164, p < 0.0001 \tag{14}$$

which can also be expressed as:

$$k20^\circ\text{C (mm/s)} = 47.04A^{6.04}B^{-0.34} \tag{15}$$

The coefficient and exponents obtained based on this database (in Eq. 15) differ with Feng et al. (2018a) (in Eq. 8), which might be due to the diversity of gradation ranges and material types involved in this study. Figure 6 shows that 84 out of 164 of the total data points lie within the ±75% bounds, and 41.5% of the points fall above the line of equality which presents over-predicted results, while 58.5% of the data points, fall beneath which give under-predictions of *K*.

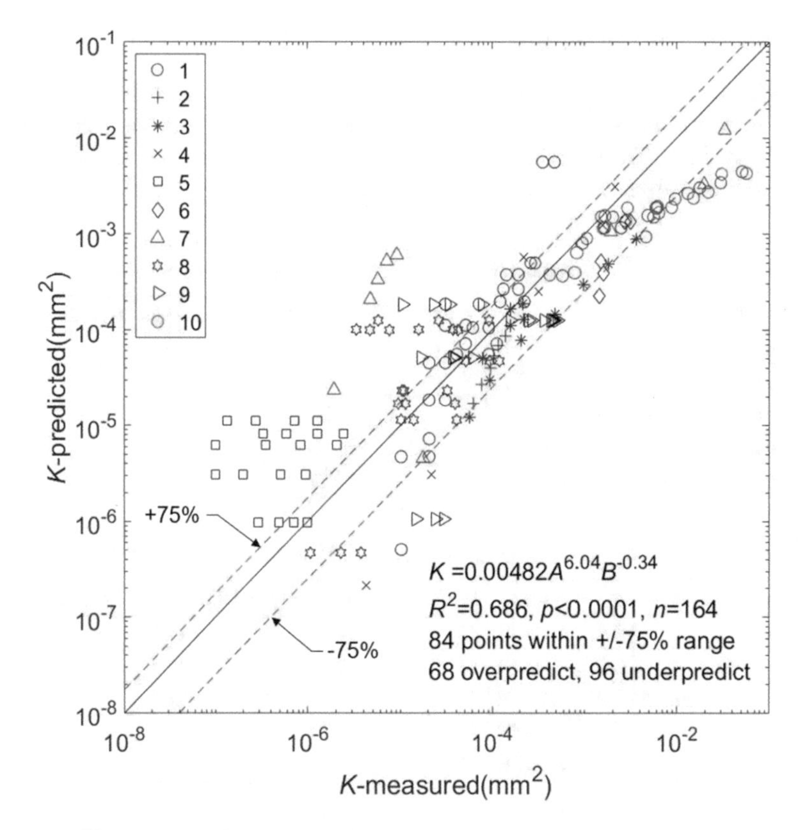

Fig. 6. *K*-predicted versus *K*-measured using grading entropy model

Summary

All five models over-predict the results for data entry 5 (Wang et al. 2017) which is possibly due to the specific type of materials used in this study (calcareous soils). Also, noticeable deviation in permeability predictions can be noticed in data source 6 (Table 6) (Xiao et al. 2013), which may be due to the fact that the gas permeameter was used. Based on the regressed results, all five of the studied models have sufficiently low *p*-values (<0.0001) and are statistically significant. The analysis results using all five models are summarised in Table 2.

Table 2. Analysis result of the complete database

n = 164	R^2	p-value	Within ±75%	Over-prediction	Under-prediction
Hazen	0.68	<0.0001	110	67	97
Shepherd	0.75	<0.0001	108	70	94
Kozeny-Carman	0.93	<0.0001	97	100	64
Chapuis	0.63	<0.0001	78	72	92
Grading entropy	0.69	<0.0001	84	68	96

3.2 Influence of Particle Size

To investigate the potential effect of particle size, the whole database is divided into two subsets, which are samples with sand component >50% and samples with sand component <50%. The classification of sand and gravel follows Craig (2004, p. 5). The results of the analysis of these two subsets are given in Tables 3 and 4.

Table 3. Analysis result of data subset: sand component >50%

n = 96	R^2	p-value	Within ±75%	Over-prediction	Under-prediction
Hazen	0.59	<0.0001	68	45	51
Shepherd	0.64	<0.0001	73	39	57
Kozeny-Carman	0.50	<0.0001	62	24	72
Chapuis	0.51	<0.0001	58	43	53
Grading entropy	0.48	<0.0001	49	57	39

Table 4. Analysis result of data subset: sand component <50%

n = 68	R^2	p-value	Within ±75%	Over-prediction	Under-prediction
Hazen	0.67	<0.0001	45	21	47
Shepherd	0.67	<0.0001	43	29	39
Kozeny-Carman	0.93	<0.0001	43	28	40
Chapuis	0.67	<0.0001	40	30	38
Grading entropy	0.74	<0.0001	45	38	30

Table 3 shows that for the data subset sand >50%, 'Shepherd' model is favoured based on the R^2 (=0.64) as well as the data points (68 out of 96) fall within ±75% prediction range, while the 'Hazen' formula gives the second highest R^2 (=0.59) and least biased predictions (45 over-predicted, 51 under-predicted).

For the data subset with sand <50% (Table 4), 'Kozeny-Carman' model does yield the highest R^2 (= 0.93) among these fives models, however, the 'grading entropy' provides the second highest R^2 (= 0.74) with most data points (45 out of 68) within ±75% prediction range and the most symmetrical prediction (38 over-predicted, 30 under-predicted) but without the need of void ratio (e) measurement.

4 Concluding Remarks

Some of the scatters shown in Fig. 2, 3, 4, 5 and 6 is inevitably due to the variety of test methods for K and e in the database. Despite this the following conclusions may be drawn:

(1) for the database with a wide range of particle sizes (the whole database), the 'Kozeny-Carman' model is favoured for the highest R^2 value, while 'Shepherd' model provides the second highest R^2 value and one of the most symmetrical predictions;

(2) for the data subset with >50% sand component in the tested mixtures, the 'Shepherd' (highest R^2, most points within ±75% margins) and 'Hazen' models (most symmetrical prediction) are preferred;
(3) for the data subset with <50% sand component, the 'Kozeny-Carman' model yields the highest R^2. However, the 'grading entropy' gives the most symmetrical prediction (38 over-predictions and 30 under-predictions) with most points fall within ±75% margins.
Further data may reveal different trends to those found in this paper and additional data to study the effect of varying testing procedures may be helpful for future research.

5 Data Availability Statement

This study has not generated new experimental data.

Acknowledgements. This first author is grateful for the financial support given by the scholarship from China Scholarship Council (CSC) under the Grant CSC No. 201708060067.

Notation List

The following notations are used in this paper (dimension given in brackets):

A = Relative base entropy;
B = Normalized entropy increment.
C_H = Hazen empirical coefficient (Length^{-1}.Time^{-1});
C_{HS} = Shepherd empirical coefficient;
$C_{K\text{-}C}$= Kozeny-Carman coefficient;
C_U= Coefficient of Uniformity, $C_U = \frac{D_{60}}{D_{10}}$;
d_{eff} = Representative particle size
D_{10}= Effective particle size, for which 10% of the soil is finer (Length);
e = Void ratio;
k = Coefficient of permeability (Length.Time^{-1});
K = Intrinsic permeability (Length^2);
n = Number of data points;
N = Number of fractions/successively doubled sieves;
p = p-value;
R^2= Coefficient of determination;
S_A= Specific surface area per unit volume of particles (Length^{-1});
S_o= Base entropy;
t = Temperature (in °C)
x_i= Relative frequency of fraction i;
γ = Unit weight (Force.Length^{-3});
μ = Dynamic viscosity (Mass.Time^{-1}.Length^{-1});
ρ = Density (Mass. Length^{-3})

References

ASTM: Standard test method for permeability of granular soils (constant head). ASTM standard D2434-68, American Society for Testing and Materials, Pennsylvania, USA (2006)

BSI: Determination of permeability by the constant-head method. BS1377-Part 5:1990. British Standards Institute, London, UK (1990)

Cabalar, A.F., Akbulut, N.: Evaluation of actual and estimated hydraulic conductivity of sands with different gradation and shape. SpringerPlus **5**(1), 820 (2016). https://doi.org/10.1186/s40064-016-2472-2

Carman, P.C.: Fluid flow through granular beds. Trans. Inst. Chem. Eng. **15**, 150–166 (1937). https://doi.org/10.1016/S0263-8762(97)80003-2

Carman, P.C.: Permeability of saturated sands, soils and clays. J. Agric. Sci. **29**(2), 262 (1939). https://doi.org/10.1017/S0021859600051789

Carrier, W.D.: Goodbye, Hazen; Hello, Kozeny-Carman. J. Geotech. Geoenviron. Eng. **129**(11), 1054–1056 (2003). https://doi.org/10.1061/(ASCE)1090-0241(2003)129:11(1054)

Chapuis, R.P.: Predicting the saturated hydraulic conductivity of sand and gravel using effective diameter and void ratio. Can. Geotech. J. **41**(5), 787–795 (2004). https://doi.org/10.1139/t04-022

Chapuis, R.P.: Predicting the saturated hydraulic conductivity of soils: a review. Bull. Eng. Geol. Env. **71**(3), 401–434 (2012). https://doi.org/10.1007/s10064-012-0418-7

Chapuis, R.P., Légaré, P.P.: A simple method for determining the surface area of fine aggregates and fillers in bituminous mixtures. In: ASTM STP, vol. 1147, pp. 177–186 (1992). https://dx.doi.org/10.1520/STP24217S

Craig, R.F.: Craig's Soil Mechanics. CRC Press, London, UK (2004)

Dolzyk, K., Chmielewska, I.: Predicting the coefficient of permeability of non-plastic soils. Soil Mech. Found. Eng. **51**(5), 213–218 (2014). https://doi.org/10.1007/s11204-014-9279-3

Feng, S.: Assessing the permeability of pavement construction materials by using grading entropy theory. M. Sc. Thesis. University of Bristol, Bristol, UK (2017)

Feng, S., Vardanega, P.J., Ibraim, E., Widyatmoko, I., Ojum, C.: Permeability assessment of some granular mixtures. Géotechnique (2018a)

Feng, S., Vardanega, P.J., Ibraim, E., Widyatmoko, I., Ojum, C.: Assessing the hydraulic conductivity of road paving materials using representative pore size and grading entropy. ce/papers **2**(2–3), 871–876 (2018b). https://doi.org/10.1002/cepa.780

Ghabchi, R., Singh, D., Zaman, M.: Laboratory evaluation of stiffness, low-temperature cracking, rutting, moisture damage, and fatigue performance of WMA mixes. Road Mater. Pavement Des. **16**(2), 334–357 (2015). https://doi.org/10.1080/14680629.2014.1000943

Goetz, R.O.: Investigation into using air in the permeability testing of granular soils. Technical report. University of Michigan, USA (1971). https://deepblue.lib.umich.edu/bitstream/handle/2027.42/5147/bac3009.0001.001.pdf?sequence=5&isAllowed=y. Accessed on 18 July 2018

Hazen, A.: Some physical properties of sands and gravels with special reference to their filtration. In 24th Annual Report of the State Board of Health of Massachusetts, pp. 539–556. Wright & Potter Printing, Boston, United States of America (1893)

Indraratna, B., Nguyen, V.T., Rujikiatkamjorn, C.: Hydraulic conductivity of saturated granular soils determined using a constriction-based technique. Can. Geotech. J. **49**(10), 607–613 (2012). https://doi.org/10.1139/T2012-016

Kandhal, P., Rickards, I.: Premature failure of asphalt overlays from stripping: case histories. Technical report NCAT 01-01 (2001). http://www.eng.auburn.edu/files/centers/ncat/reports/2001/rep01-01.pdf. Accessed 18 July 2018

Kozeny, J.: Über kapillare leitung des wassers im boden: (aufstieg, versickerung und anwendung auf die bewässerung). Hölder-Pichler-Tempsky (1927, in German)

Lőrincz, J., et al.: Grading entropy variation due to Soil crushing. Int. J. Geomech. **5**(4), 311–319 (2005). https://doi.org/10.1061/(ASCE)1532-3641(2005)5:4(311)

Mallick, R.B., El-Korchi, T.: Pavement Engineering: Principles and Practice. CRC Press, Boca Raton, USA (2008)

Mavis, F.T., Wilsey, E.F.: A Study of the permeability of sand. University of Iowa, USA (1936). https://ir.uiowa.edu/cgi/viewcontent.cgi?article=1007&context=uisie. Accessed on 18 July 2018

Morris, D.A., Johnson, A.I.: Summary of Hydrologic and Physical Properties of Rock and Soil Materials, as Analyzed by the Hydrologic Laboratory of the U.S. Geological Survey. United State Department of the Interior. Washington, USA (1967). https://pubs.usgs.gov/wsp/1839d/report.pdf. Accessed on 18 July 2018

Shepherd, R.G.: Correlations of permeability and grain size. Ground Water **27**(5), 633–638 (1989). https://doi.org/10.1111/j.1745-6584.1989.tb00476.x

Singh, V.P.: Entropy Theory in Hydraulic Engineering. American Society of Civil Engineers, Reston, VA (2014). https://doi.org/10.1061/9780784412725

Thom, N.: Pothole formation: experiments and theory. Asph. Prof. (60), 22–25 (2014)

Vardanega, P.J., Feng, S., Shepheard, C.J.: Some recent research on the hydraulic conductivity of road materials. In: Loizos, A., Al-Qadi, I., Scarpas, T. (eds.) Bearing Capacity of Roads, Railways and Airfields. Proceedings of the 10th International Conference on the Bearing Capacity of Roads, Railways and Airfields (BCRRA 2017), Athens, Greece, June 28–30, 2017, pp. 135–142. Taylor & Francis, London, UK (2017). (full-paper on USB)

Wang, X.Z., Wang, X., Chen, J., Wang, R., Hu, M., Meng, Q.: Experimental study on permeability characteristics of calcareous soil. Bull. Eng. Geol. Env. (2017). https://doi.org/10.1007/s10064-017-1104-6

Xiao, Y., Tutumluer, E., Moaveni, M.: In-situ hydraulic properties of unbound aggregate layers measured using gas permeameter test (GPT) device. In: Airfield and Highway Pavement 2013, pp. 1370–1385. American Society of Civil Engineers, Reston, VA (2013). https://doi.org/10.1061/9780784413005.116

Yin, J., Hachiya, Y.: Permeability of drainage base course materials a laboratory tests. 土木学会铺装工学论文集 **3**, 175–182 (1998)

Virtual Reality and Neural Networks for Exploiting Geotechnical Data

Silvia García[1(✉)], Paulina Trejo[1], Alberto García[2], and César Dumas[3]

[1] Geotechnical Department, Instituto de Ingeniería UNAM, Mexico City, Mexico
sgab@pumas.iingen.unam.mx
[2] Head of GEOIntelligent Systems Laboratory, Mexico City, Mexico
[3] Head of the Geotechnical Laboratory, Comisión Federal de Electricidad, Mexico City, Mexico

Abstract. Design of foundations for large-scale civil works naturally involves soil characterization over considerable volumes. The 3D-interpretation of properties where only scarce geotechnical data is available is crucial for deriving effective and safe engineering decisions. Because of the ever-increasing cost of site investigation, it is neither practical nor economical to acquire geotechnical data at each point of interest for a complete definition of soils behavior. This situation makes it necessary to explore spatial-variability modeling alternatives that can manage limited geo-information. In this paper, a dynamic-neural procedure is developed for describing spatial relations between a set of geo-parameters easy-to-obtain. Once the network is finished, this topology is used to expand the small initial set of values into millions of computer-generated measurements. The massive database is incorporated into a Virtual Reality engine that facilitates the intuitively visual understanding of geo-information, permits to present all relevant data in a comprehensible format for decision making and provides a way to reduce very complex and diverse data sets into the essential elements without loss of data quality.

1 Introduction

The spatial variability of soil properties poses a major challenge to geotechnical engineering. In order to design a secure structure, it is necessary to compile the largest quantity and the best quality of soil information. Field investigations imply a well-planned program of boring, sampling and experimental testing as well as an adequate distribution of monitors. Then from the interpretation of the results, appropriate models for mimicking the in-situ soil behavior must be developed with the purpose of coupling them with numerical tools. It is worth stressing the fact that regardless the detail with which the geotechnical field studies are planned and existing information considered, it is unlikely that all the geological details be brought out and hence critical information might not be accounted for in the design analyses.

Because of these uncertainties and the liability costs of performing exhaustive geo-investigations, engineers have resorted to available interpolation techniques such as the kriging and co-kriging procedures. However, it should be understood that most of the

S. Hemeda and M. Bouassida (Eds.): GeoMEast 2018, SUCI, pp. 14–30, 2019.
https://doi.org/10.1007/978-3-030-01941-9_2

interpolation schemes require a significant amount of input data in order to yield acceptable estimations of the design parameters. Accordingly, to overcome this shortcoming alternate interpolation methods, such as Neural Networks NNs, should be considered.

In this paper, a recurrent neural network RNN is used to study a set of geotechnical parameters in a geographical context. The application example is on Lake Texcoco clays. It is worth noticing that this kind of spatial analysis can handle uncertain, vague and incomplete/redundant data. The spatial-neural model contains the relations between the spatial patterns of the stratigraphy without restrictive assumptions or excessive physical simplifications. In addition, the neural capabilities for interpreting soil characteristics in large sites, from a relatively small amount of data, are shown.

The RNN model is used to feed a Virtual Reality VR engine. VR is an emerging technology to simulate the real world on the computer and is being applied to many fields of industries like Civil Engineering. The uses of Virtual Reality not only make us transverse the time and space to *feel* a planning task but also avoid a quickly understanding on any complex project. Physical information is directly displayed as 3-dimensional objects in the ground model. Numerical and text information is dynamically linked to these objects and accessed by "clicking" on them, facilitating easy access to and integration of laboratory and *in-situ* data, all within one application. Based on the results, it can be concluded that VR technology provides an intuitive, multi-dimensional information system which can simultaneously display any information that can be collected and stored by a computer.

2 Architecture of the Recurrent Neural Network

The standard feedforward neural network, or multilayer perceptron MLP, is the best-known member of the family of many types of neural networks [1]. Feed-forward neural networks have been applied in tasks of prediction and classification of data for many years [2, 3]. The dynamic NN, or neural networks for temporal processing, extend the feedforward networks with the capability of dynamic operation. This means that the NN behavior depends not only on the current input as in feedforward networks but also on previous operations of the network.

The RNN, as networks for temporal processing, gains knowledge through recurrent connections where the neuron outputs are fed back into the network as additional inputs [4]. The fundamental feature of a RNN is that the network contains at least one feedback connection, so that activation can flow around in a loop. This enables the networks to perform temporal processing and learn sequences (e.g., perform sequence recognition/reproduction or temporal association/prediction). The learning capability of the network can be achieved by gradient descent procedures, similar to those used to derive the backpropagation algorithm for feedforward networks [5].

The network consists of a static layer, which generally has a higher number of neurons compared to the number of state variables of the system, from which the output is directed to an adder, where it is subtracted from the previous value of the variable Z_i, identified by the system. From this operation, the derivatives of each of the i state variables identified by the system are generated. The dynamic recurrent multilayer network described in Eq. 1 can identify the behavior of an autonomous system ($u = 0$, Eq. 2):

$$\frac{d}{dt}z = \bar{f}(z) = Ax + \omega\sigma(Tz) \quad (1)$$

$$\frac{d}{dt}x = f(x) = Ax + f_o(x) \quad (2)$$

Here

$$\begin{aligned} &x, z \in R^n, A \in R^{nxn}, f(x) : R^n \rightarrow R^n, \bar{f}(z) : R^n \rightarrow R^n, \mathrm{w} \in R^{nxN}, \\ &T \in R^{nxn}, \sigma(z) = [\sigma(z_1), \sigma(z_2), \ldots, \sigma(z_n)] \end{aligned} \quad (3)$$

the transfer function $\sigma(\theta) = tansig(\theta)$, n is the number of state variables of the system, N is the number of neurons in the hidden layer, and $f_o(x)$ is the estimated $f(x)$. According to Haykin [1], without loss of generality, if the source is assumed to be an equilibrium point, the system (Eq. 2) will be identified with the network (Eq. 1) about its attraction region and guarantees that the error in the approximation $e(t)$ is limited. Mathematical and algorithm details of the proposed procedure are included in [6].

2.1 Learning Rules

The static stage of the dynamic recurrent multilayer network is usually trained with a backpropagation algorithm. These algorithms are widely described in the vast literature, for example in [7, 8]. The training patterns of the static layer are different combinations of values of the state variables, and the target patterns are given by the sum of each state variable with their corresponding derivative, as shown in Fig. 1. The network is trained after the structure of Eq. 4:

$$\begin{aligned} \frac{d}{dt}\begin{bmatrix} z_1 \\ z_2 \\ \cdot \\ z_n \end{bmatrix} &= \begin{bmatrix} -z_1 \\ -z_2 \\ \cdot \\ -z_n \end{bmatrix} + \begin{bmatrix} w_{11} & w_{12} & \ldots & w_{1n} \\ w_{21} & w_{22} & \ldots & w_{2n} \\ w_{n1} & w_{n2} & \ldots & w_{nn} \end{bmatrix} \\ &\times \begin{bmatrix} \sigma(t_{11}z_1 + t_{12}z_2 + \ldots + t_{1n}z_n) \\ \sigma(t_{21}z_2 + t_{22}z_2 + \ldots + t_{2n}z_n) \\ \cdot \\ \sigma(t_{n1}z_1 + t_{n2}z_2 + \ldots + t_{nn}z_n) \end{bmatrix} \end{aligned} \quad (4)$$

Here t_{ij} are expected values of the variable. To ensure that the network has identified the system dynamics, the Jacobian of the network at the source (Eq. 4) should have values very close to those of the system that has been approximated.

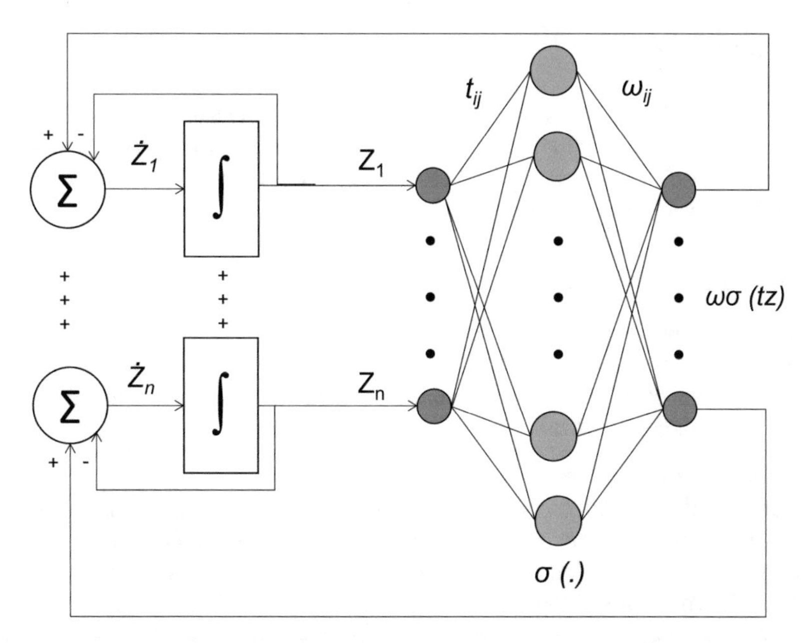

Fig. 1 Dynamic multilayer networks. z_i is the output of the static layer i, $\dot{Z}$, is the derivative with respect to time of Z_i, t_{ij} is the expected value, ω_{ij} is the weight and σ is the transfer function.

$$J_M = -I_n + WT \tag{5}$$

where J_M is the Jacobian, In is the identity matrix of dimension n, W is the weights matrix, and $T = \sigma(t_{ij}\,z_j)$. The dynamic multilayer network of Fig. 1 can be transformed into a dynamic network (Hopfield type) by means of the following linear transformation:

$$X = Tz \; and \; \frac{dx}{dt} = T\frac{dz}{dt} \tag{6}$$

Generally, the T matrix is square, but if it is not, the transformation is performed by means of the generalized inverse. The transformed network will have the structure

$$\frac{d}{dt}x = -INX + TW\sigma(x) \tag{7}$$

Here, IN is the identity matrix of dimension N, and the transformation (Eq. 3) extends the dynamic multilayer network (Eq. 3) into the dynamic recur-rent Hopfield network (Eq. 5). An algorithm of the proposed RNN procedure, which includes the steps implicit in the shown Equations was developed using the tools of the Mathworks website.

3 Virtual Reality

Virtual Reality VR creates a portal through which a person can enter a new, immersive, human-designed world. Unlike other interfaces, in which the user views a screen while remaining otherwise tethered to their contextual physical world, VR allows users to experience a whole different universe.

3.1 How Does VR Work?

VR systems induce us to believe we belong to a specific environment. To reach this, the system has to mislead the parts of the brain that perceive vision, motion and other senses. Any competent VR system should have a variety of components to build this illusion, most important of them being [9]:

Head Mounted Display

The most significant part of any VR system is the head-mounted display (HMD), responsible for trapping our vision. To understand how a HMD operates, first it is necessary to describe how human vision works. Basically, each of our eyes captures 2D images of the environment we are looking at, but our brain is the processor that does the trickery to perceive what we see as 3D using many other information like spatial audio, interaction with the environment, previous experiences etc. [10]. VR systems use the principle of stereoscopic vision to simulate the perception of depth and 3D structures. To achieve this, in VR separate images for each eye are generated, one slightly offset from the other, to simulate parallax like the diagram in Fig. 2.

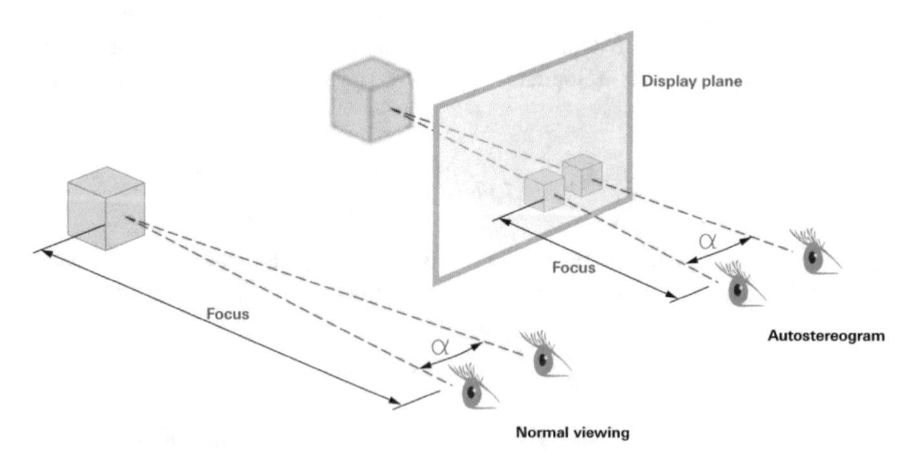

Fig. 2 The VR system renders a stereoscopic image on the display of the HMD.

Display

The VR system renders a stereoscopic image on the display of the HMD with a minimum frame rate of 60 frames per second, to avoid any perceived lag that might break the illusion or, worse, lead to the nausea that is often associated with poorly performing VR.

Optics
If you ever try to hold a stereoscopic image close to your eyes (typically at the distance HMDs keep the screen), you will not be able to make out what you are seeing because your eyes won't be able to converge them on the same plane (right image on the above diagram). The HMD includes lenses that augment our eyes to converge the images, so the VR system renders something like the image in Fig. 3 on the display and the lenses converges and corrects distortions in order our brain finally perceives something "real" with a sense of depth.

Fig. 3 A pair of lenses that augments our eyes to converge the images, corrects distortions and finally perceives with a sense of depth (shared by D Coetzee, http://blog.dsky.co).

Motion Tracking
After display, the next essential trick for making the brain believe it is in another place is to track movements of the head and update the rendered scene without any lag (Fig. 4). The VR system has to do this within 16.67 ms to maintain 60 frames per second. This metric is known as Motion to Photon (MTP) latency. For details about MTP latency, consult [11].

Usually an array of sensors like gyroscopes, accelerometers etc. are used to track the movement of our head and the information is passed to the computing platform to update the rendered images accordingly. Apart from head tracking, advanced VR systems targeting for better immersion also tracks the position of the user in the real world.

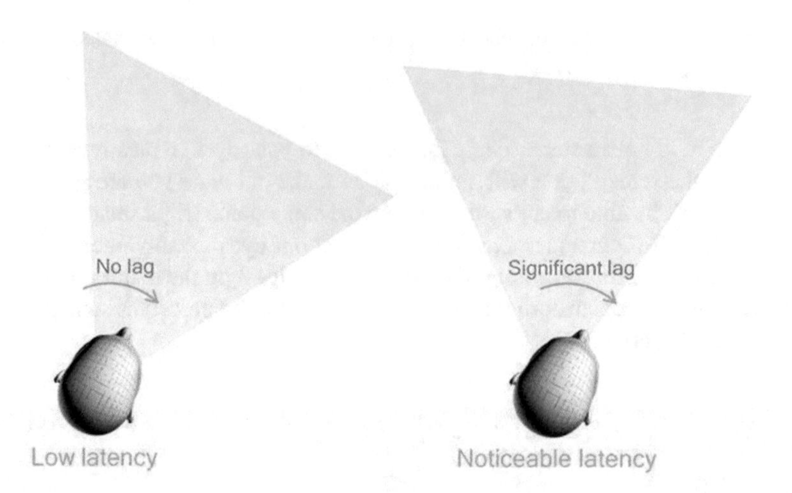

Fig. 4 The motion tracking and the trick for making the brain believe it is in another place.

Interaction Devices/Controllers
To increase the degree of immersion a VR system has to allow the user to interact with the virtual environment. This asks for specialized input devices, which can as simple as a magnetic button in *Google Cardboard* or as advanced as hand and body-tracking sensors that can recognize motion and gestures (like *Leap Motion*, *HTC Vive body* tracking system, etc.).

Computing Platforms
Finally a computing platform, which have to do all the above heavy processing very fast. Usually VR systems use either PC/Console or smartphones as the computing platform.

3.2 Practical Applications of VR

Virtual reality was first adopted as a gaming platform that could allow gamers to go further than the existing screen-interface options allowed. However, the potential of VR technology applications in many other industries was clear, and funders with large-scale commercial goals began to invest in its future. The first of the productive activities that showed the product and benefited economically from its capabilities was the filmmaking. The cinematographers experimented using 3D cameras to capture a 360-degree experience while the music producers are exploring the possibility of giving fans virtual front-row seats at performances. News organizations such as the New York Times, ABC and Huffington Post have spent millions to bring VR capacity to their breaking news stories.

Some of the uses of VR by the medical community are the exposure therapy to conquer phobias and pre-procedure imaging, which enables it to better strategize the approach and minimize surgery time. The applications of VR to Civil and Architectural Engineering projects have concentrated on visual simulation of the construction process, management of the design information, geometrical representations of the civil works together with scheduling data (the 4D = 3D + time), among others. Undoubtedly, one of the engineering branches that most successful VR-products have developed is geology, with noteworthy studios for 3D data visualization, geo-interpretation software for field campaigns, digital outcrop models with multiple attributes from any point cloud type dataset, among others.

4 Geo-Environment: Lake-Texcoco Clays

The lakebed soils of downtown Mexico City have been extensively described [12, 13]. However, the Lake Texcoco deposits (outside the city built up area, see Fig. 5) had less attention until the mega-projects programmed for the area faced these difficult soils, reactivating intense and deep geo-investigations in the region.

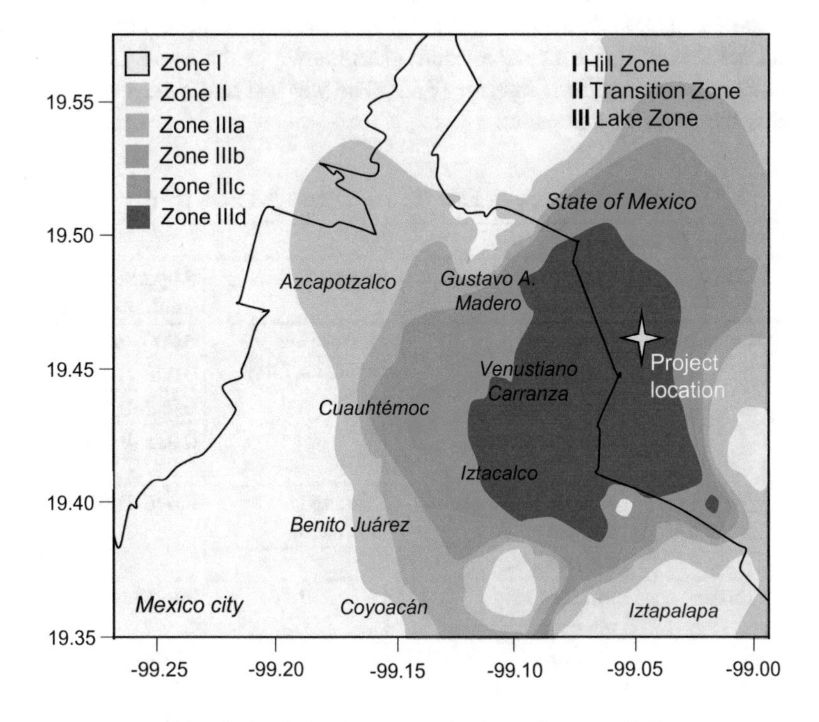

Fig. 5 Analysis sites over seismic zoning map [19].

The data were obtained from a private source so it is not possible to publish details of the exact location or the total of the shared material. For the purposes of this document, it is sufficient using some of the information to show the advantages of the RNN-VR techniques for analyzing an area in the ex-Lake ($\approx$80 km^2).

4.1 Depositional Environment

The Basin of Mexico occupies an area of 10,000 km^2 and is a predominately a flat lacustrine plain. The Basin was open until about 700,000 years ago, during the Pleistocene epoch, when volcanic activity caused the creation of the Chichinutzin Range. Surrounding the Basin are ranges of alluvial fans and debris flows (lahars) interbedded with volcanic pumice and ash. This volcanic-sedimentary complex is known as the Tarango Formation.

Lake Texcoco was formed within the Basin by periods of glaciation and persistent rains within the last 100,000 years. As the surrounding mountains gradually eroded, the fine and ultra-fine volcanic ash particles were transported within the basin. Subsequent volcanic eruptions created dense ash clouds that settled onto the lake surface by rainfall. A feature of the lacustrine deposits is the ubiquitous presence of diatoms, the skeletal remains of organisms that thrived in the nutrient-rich lakebed [14].

Table 1 summarizes the general geology of Lake Texcoco. The Holocene Clays comprise the *Formacion Arcillosa Superior* (FAS) that is typically 25 to 30 m thick, a 2 to 5 m thickness of dense sands/volcanic glass known as the *Capa Dura* (CD) followed by *Formacion Arcillosa Inferior* (FAI). The weakest and softest material is the FAS and is the focus of this paper.

Table 1. Geology of Lake Texcoco (modified from [14])

Geology of Lake Texcoco				
Depth (m)	Period	Epoch	Formation	Approximate age (millions of years)
0–53	Quaternary	Holocene	Lacustrine deposits. Clay	0.000–0.008
53–59		Late pleistocene		0.008–0.012
59–64		Late pleistocene		0.012–0.013
64–180		Late pleistocene		0.013–0.046
180–505		Middle pleistocene to late pliocene	Tarango formation	0.046–0.800
505		Refractor A		
505–814	Tertiary	Middle/Early pliocene		8.0–13.0
814–1030		Miocene	Huatepec	13.0–21.0
1030–1125		Early miocene-late oligocene		21.0–24.0
1125–1437		Late oligocene		24.0–29.0
1450		Refractor B		
1437–2065		Middle Oligocene-middle eocene	Balsas formation	29.0–(?)

4.2 Description of the Geotechnical Properties of FAS

The following properties at the site were obtained from testing by [15, 16]:

Pore water salinity: typically 45 g/L;
Clay fraction: typically 50 to 60%;
Uncorrected water content: 150 to 350%;
Liquid limit: 300% reducing to 150% with depth;
Plastic limit: 40% to 100%;
Unit weight: 12 kN/m^3;
Void ratio: 4 to 8; and
ϕ' (from CIU testing, valid up to shear strains of 10%): 40°, with standard deviation (S.D.) of 7°.

High ϕ' values are typical of Lake Texcoco clays and these are usually ascribed to the effect of diatoms. Relations between diatom content and engineering properties can be consulted in [17, 18].

In the last 50 years the ground surface the center of Lake Texcoco has settled at least 8 m because of the pumping from the lower aquifers. One effect of pumping groundwater has been the lowering of the piezometric profile to less than hydrostatic within some soil units. The resulting consolidation of the soils in the *Formacion Tarango* and above has led to regional subsidence as high as 35 cm/year [14].

The data used in this investigation was determined from laboratory and *in-situ* tests. The ≈80 km^2 of examined area had a maximum depth of exploration of 90 m (Fig. 6). From the results of the field exploration, the standard penetration tests (SPT) and the cone penetration tests (CPT) were selected to show the procedure for modeling multi-parametric spatial variability, and the natural water content (W%) is included as a support parameter. Add W% permit to dissolve the possible contradictions between strength properties of FAS materials measured *in situ* and to refine the broad stratigraphic descriptions.

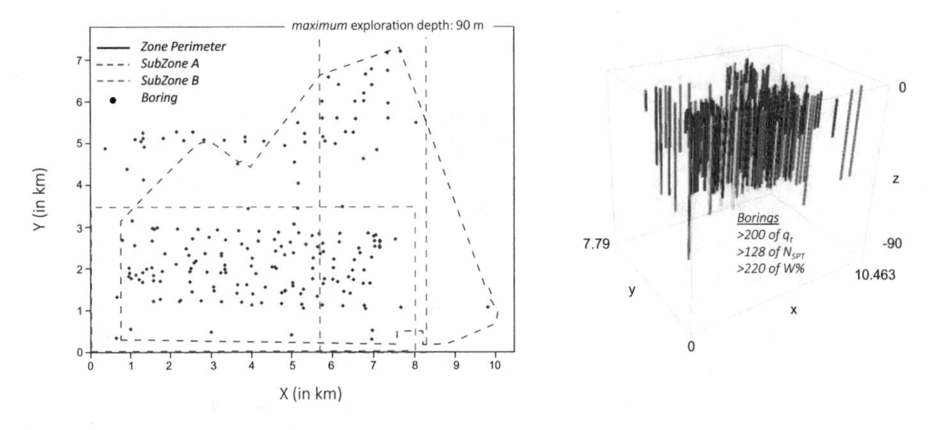

Fig. 6 Geometry of the volume involved in the large-scale project, covers approximately 70 km square area.

The SPT results are presented to describe how this easy-to-obtain parameter (blow counts N_{SPT}) can be exploited for giving significance to engineering geological site characterizations. Because of the connection between the interpreted results and the relative density, shear strength of soils and the bearing capacity, it is hypothesized that N_{SPT} can be directly modeled with the cone penetration test (cone resistance q_t) to delineate soil stratigraphy.

5 VR-Neural Exploration

The first step is constructing a neural-estimator for defining the spatial variation of the support parameter W%. To generate water contents at any location inside the studied volume, a RNN is trained in a first stage, using the W% determined during the exploration campaign. 40,000 tested samples from 226 soil borings conform the W%-database. Taking advantage of the dynamic-methods abilities, a virtual population of values is obtained (W% neuro-estimated in locations where no samples were taken for the lab, sufficiently close to the real ones so that the Tobler's Law can be enforced) and used for updating the RNN model.

When virtual boring situations (to a spatial distance from a training boring of 0.1 km and 0.1 m on the plane and depth, respectively) are estimated, these "new" W% are placed in the queue of the training file. Each time the virtual W% are presented to the network, its weights are adjusted, so the initial RNN is retrained as many times as numbers of estimations are needed for covering the studied space.

After thousands of estimates in virtual stations, the volume under study is fully covered and the training is stopped, considering the spatial model terminated. Details about the step-by-step procedure applied to real-world observations, within a geographic space, can be consulted in [18].

An example of the neural spatial distribution for W%, is shown in Fig. 7. Based on the estimated properties for the millions of voxels (14 millions for the volume of $\approx$7.1 km^3), calculated with the RNN, it can be precisely defined the 3D situations of maximum and minimum values of the water contents. For the selected subzone, among the remarkable findings of this model is the location of the material, from the surface, which presents the highest or anomalous levels of W%. From the spatial distribution, it can be detected those zones that could be directly related to the stratigraphic sequences, allowing these to be better and more precisely detailed.

It is distinguished a layer of material, from the surface and up to $\approx$25 m, which presents the highest levels of W% reaching up to 450%. From this stratum, a drastic change of behavior is recognized, crossing the CD, with subsequent clays that have W % far below the average ($\approx$250%).

Once the W% display is defined, the procedure discussed above is applied to develop the neural-models for the N_{SPT} and q_t spatial variations. Unlike the first model, that of the support parameter, for training these neural networks, each $N_{SPT} = f(X,Y,Z)$, $q_t = f(X,Y,Z)$ is coupled with its respective W% and they are arranged as the inputs [X,Y,Z, W%] – output [N_{SPT}/q_t] file. As was done for the water content in the iterative procedure, for each resistance measurement, the spatial distances for virtual stations are set at 0.1 km and 0.1 m. These models resulted in more complex architectures,

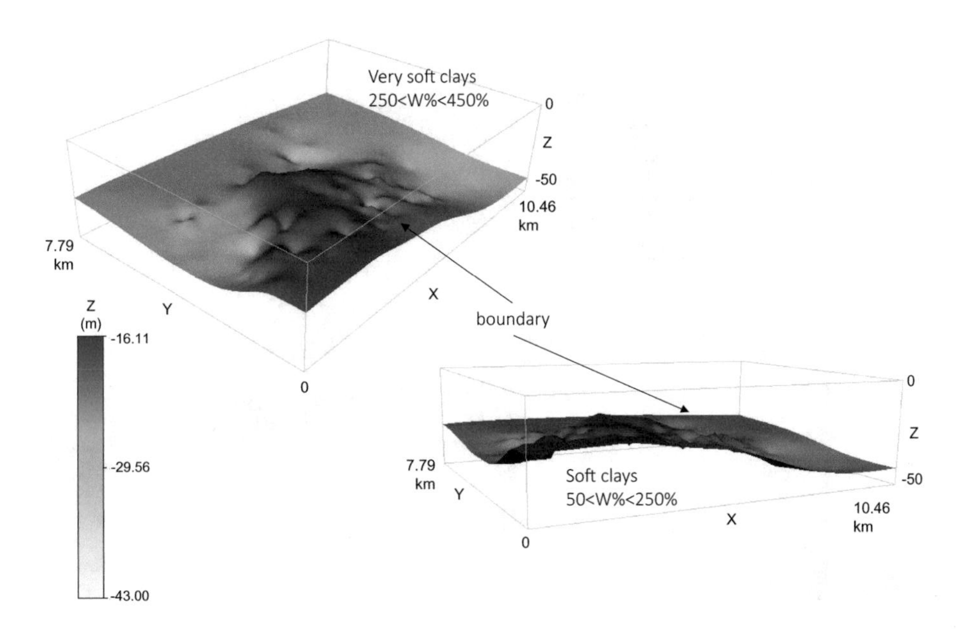

Fig. 7 Spatial distribution for W%: boundary between materials with maximum values of W% (250 to 450%) and those that have W% below the media (250%).

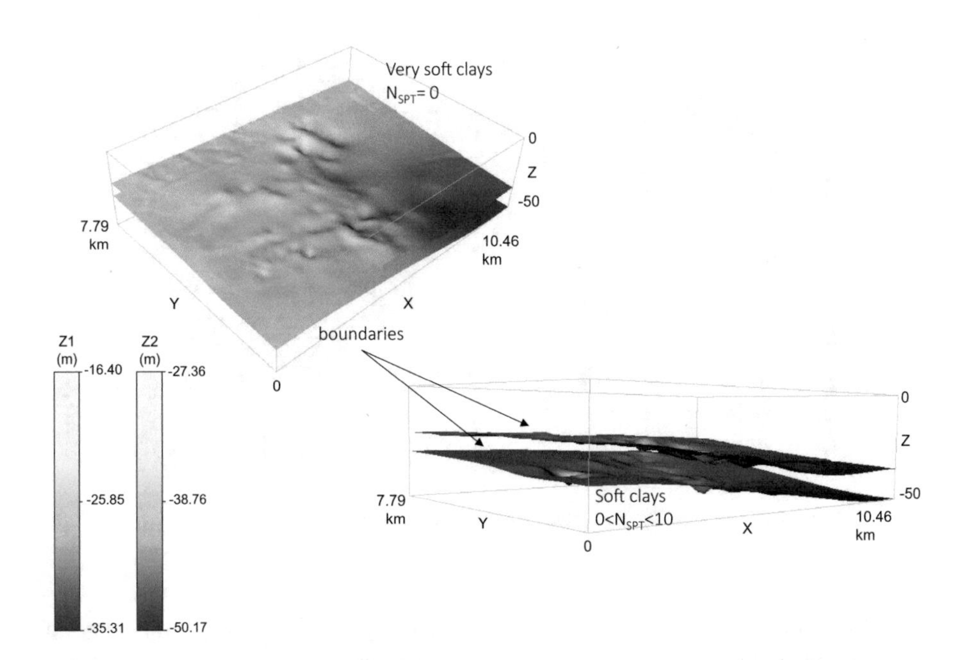

Fig. 8 Spatial distribution for N_{SPT}: boundary between materials with $N_{SPT} \approx 0$ and those whose rigidity did not allow the test to be developed (very rigid).

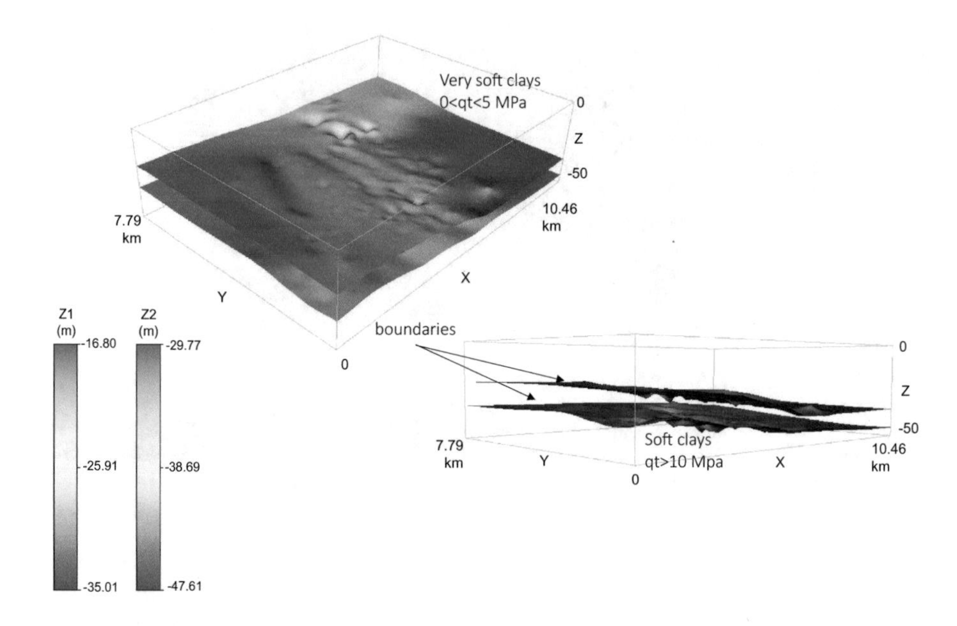

Fig. 9 Spatial distribution for q_t: boundary between materials with $q_t \approx 0$ (<5 MPa) and those with values above 10 MPa.

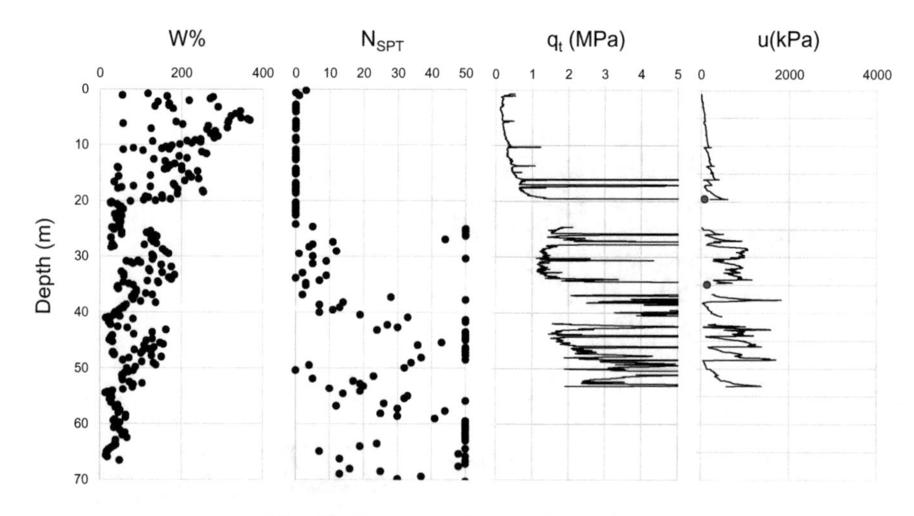

Fig. 10 Examples of properties profiles.

requiring more than seven times the number of nodes that were needed to model the W %. To achieve the same level of accuracy (precision in the prediction), the number of iterations, computation time, was significantly increased with respect to the first model.

As with the W% display, in Figs. 8 and 9 subzones of the neural spatial distributions (for N_{SPT} and q_t) are shown. It is very important to point out that each voxel in these volumes corresponds to those used to build the 3D of the water content, thus, the geo-information can be referenced at each point in the space (X,Y,Z) with a complete set of properties. For the construction of the N_{SPT} display, each cube is not "filled" with color since all the profiles (see Fig. 10) have a clear sequence of constant values (near to zero blows) what produces large regions with practically the same value. Changes in N_{SPT} are observed only once the CD has been traversed. Using the RNN a layer of material whose rigidity is comparable to that of the CD was detected.

With the superimposed voxels of N_{SPT} and q_t the position of the CD is corroborated. Once again, the almost constant values of the property produces sub-volumes

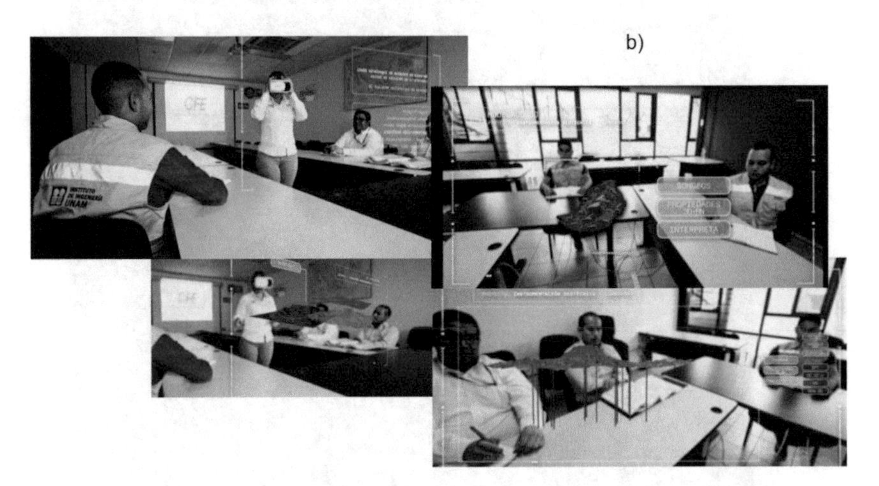

Fig. 11 The VR system situated in a dedicated room in CFE (in the Federal Electricity Commission), a) laboratory administration, b) team work.

with practically nule resistance (there is hardly an increase in resistance as the depth advances). However, the maximum and minimum values of resistance are defined for each position. Exploiting the advantageous neural-capabilities, the hypothesis of almost horizontal depositions without anomalies is annulled and the peculiar characteristics of the layers are easily detected and displayed.

The product presented to the developer of the civil project is a 3D observation camera that can rotate at the user's will and allows to obtain cuts and plants to accurately define the presence and properties of each layer of material.

6 A VR-Engine

Combining the powerful data mechanism created with the RNNs with a virtual reality engine enables to visualize, analyze and share this complex geo-scientific dataset seamlessly in an immersive 3D, real-time environment. The approach proposed here is different from other VR and GIS mapping because the interest is focused on viewing all of the data, all of the time, in 3D, at the highest level of detail.

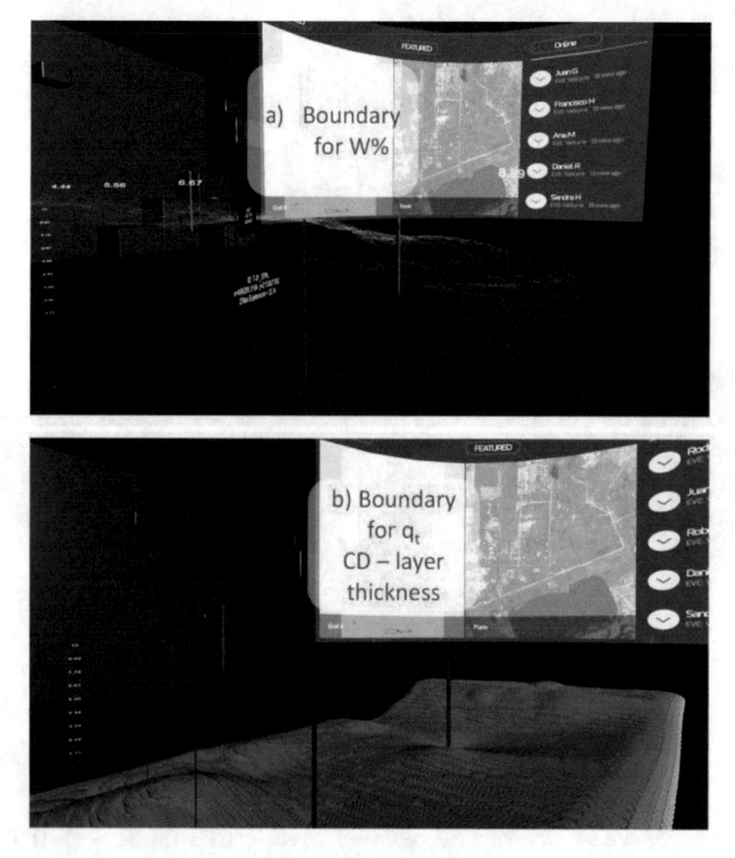

Fig. 12 Discussion arena, a) description of W% , b) cone penetration - variation of q_t; in these rooms you can attract diverse test results as well as remote interaction with other work teams.

The idea in the VR engine is to manage the design, planning, and execution stages seeing exactly what the engineers are looking for. The visualization of information is vital for understanding, discussing, planning and making right decisions. The VR system is situated inside a dedicated room within in the Geotechnical Department of CFE (in the Federal Electricity Commission) (Fig. 11), but also has can be exploited in personal gadgets which enable stereoscopic vision.

The RNN-VR system was designed to be, in addition to a powerful visualization tool, an environment to manipulate the numbers and models "from within". This means that the engineers team experienced a highly immersive calculation process. Inside the neural 3D-body the user can select a point, ask for its registered characteristics and estimate additional properties with the highly detailed RNN-models (Fig. 12).

7 Conclusions

Conventional interpolating estimators require a great number of measurements available, what is generally impractical. In this investigation, a RNN is presented as a very efficient multiparametric-tool for the interpretation of available geotechnical information. The neural technique can be used to integrate systematically the results of soil monitoring and exploration, handling the inherent in situ testing uncertainty and taking advantage of broad stratigraphic descriptions.

The most obvious factor affecting spatial variation is the relative position be-tween the monitors and the selection of samples to be teste in laboratory. To obtain the maximum benefit at minimum cost, meaning that the number of evaluations must be just sufficient to ensure continuity, without costing more than necessary, the inevitable question "what is the optimum stations spacing?" arises. The answer can be obtained easily, efficiently, and truthfully with the neural technology proposed. Following the results presented it can be detected the best positions for new exploration campaigns.

The VR-engine applied to geotechnical exploration is based on an extensive-scale product. By this introduction it is recognized that VR permits a quick perception to complex projects, before project is carried out, and the intuitively visual understanding permits to minimize interpretation errors. The product is designed for individual observation but its greatest potential is obtained when it is operated in interaction with work teams. A good communication among the participants of a project can be reached with a good education effect.

References

1. Haykin, S.: Neural Networks and Learning Machines, 3rd edn. Pearson Prentice Hall, New Jersey (1999)
2. Egmont-Petersen, M., de Ridder, D., Handels, H.: Image processing with neural networks-a review. Pattern Recognit. Soc. **35**, 2279–2301 (2002)
3. Theodoridis, S., Koutroumbas, K.: Pattern Recognition, 4th edn. Academic Press, Massachusetts (2008)

4. Graves, A., Liwicki, M., Fernández, S.: A novel connectionist system for unconstrained handwriting recognition. IEEE Trans. Pattern Anal. Mach. Intell. **31**(5), 855–868 (2009)
5. Hinton, G., Osindero, S., Teh, Y.-W.: a fast learning algorithm for deep belief nets. Neural Comput. **18**, 1527–1554 (2006)
6. Hochreiter, S., Younger, A.S., Conwell, P.R.: Learning to learn using gradient descent. In: Artificial Neural Networks-ICANN International Conference Proceedings, Austria, pp. 87–94 (2001)
7. Serrano, M.Á., Boguñá, M., Vespignani, A.: Extracting the multiscale backbone of complex weighted networks. Phys. Soc. **106**(16), 6483–6488 (2009)
8. Aguayo, J. E.: Neotectónica de facies sedimentarias cuaternarias en el suoreste del Golfo de México, dentro del marco tectónico-estratigráfico regional evolutivo del Sur de México. Ingeniería: Investigación y Tecnología **VI**(1), 19–45 (2005)
9. Das, N.: Working in Graphics & VR at Samsung R&D India. An-swered 12 July 2017 in www.quora.com
10. Di Girolamo, J.: The Big Book of Family Eye Care, pp. 96–97. Basic Health Publications, Inc., Laguna Beach (2011). 111
11. OCULUS, Dec 2012. https://www.oculus.com/blog/details-on-new-display-for-developer-kits/
12. Santoyo, E., Ovando, E., Mooser, F., León, E.: Síntesis geotécnica de la cuenca del Valle de México. TGC geotecnia S.A. de C.V., México, D.F., 171 p. (2005)
13. Diaz-Rodriguez J.A.: Characterization and engineering properties of Mexico City lacustrine soils. In: Tan, T.S., et al. (eds.) Characterization and engineering properties of natural soils, vol. 1, pp. 725–755. Swets & Zeitlinger, Lisse (2003)
14. O'Riordan, N., Canavate, A., Ciruela, F.: The stiffness and strength of saltwater Lake Texcoco clays, Mexico City. In: Proceedings of the 19th International Conference on Soil Mechanics and Geotechnical Engineering, Seoul, pp. 1067–1070 (2017)
15. Fugro: Informe de pruebas de laboratorio en muestras de suelo en el área destinada al nuevo Aeropuerto de la Ciudad de México (2016)
16. IE Ingeniería Experimental: Informe de investigación Geotécnica del sub-suelo del Ex-Lago de Texcoco para el Nuevo Aeropuerto Internacional de la Ciudad de México (2015)
17. Diaz-Rodriguez, J.A., Santamarina, J.C.: Strain-rate effects in Mexico City soil. J. Geotech. Geoenviron. Eng. **135**(2), 300–305 (2009). Tech Note. ASCE
18. García, S., Alcántara, L.: A neurogenetic model for determining spatially variation of PGA. J. Earthq. Eng. (2018, submitted)
19. GDF: Normas técnicas complementarias para diseño por sismo (2004)

Validation of Compression Index Approximations Using Soil Liquid Limit

Amir Al-Khafaji(✉), Abbey Buehler, and Ethan Druszkowski

Department of Civil Engineering and Construction, Bradley University, Illinois, USA
amir@fsmail.bradley.edu

Abstract. The list of published empirical models for approximating the compression index of fine-grained soils in terms of its liquid limit continues to grow. Many such expressions are justified based on high correlation coefficients using limited test data sets. Significant measures such as standard errors are often omitted. Some published models are applicable to fine-grained soils, organic soils or all soils. Most of the published models are based on traditional linear regression which ignores data outliers. Artificial Neural Networks (ANN) regression are being used to develop more reliable models but often ignores data outliers. Although ANN is a valuable tool, it is data driven and the quality and quantity of the data impact developed models. Over the past five years, significant data was collected from published articles from reputable journals and conferences. A total of 1906 data points relating soil compression index to its liquid limit were analyzed using Robust Bi-Square regression to reduce the impact of data outliers. Published models rely on traditional regression analysis. The results indicate several published data sets involving liquid limits of less than 16 while others included data that plotted above the U-line in the Plasticity Chart. Such data points are impossible to exist in engineering practice. Using MATLAB and the Robust Bi-Square regression, the authors developed new regression models with high correlations and low standard errors that relate the compression index of fine-grained soils to its liquid limit. The proposed models are more reliable than those developed using traditional regression technique which ignore data outliers and often involve incorrect data. The proposed Robust regression models were compared with published empirical approximations that are based on traditional regression methods to establish validity and reliability.

Keywords: Regression · Empirical · U-line · ANN · Robust · Coefficient Compression · Index · Liquid limit · MATLAB

1 Introduction

The calculation of consolidation settlement of foundations placed on normally consolidated fine-grained deposits requires knowledge of soil compression index, C_c. This requires standardized consolidation tests which are both time consuming and expensive. The number of tests required depends on the variability of soil profile and loads involved. The development of empirical equations for estimating the compression

S. Hemeda and M. Bouassida (Eds.): GeoMEast 2018, SUCI, pp. 31–41, 2019.
https://doi.org/10.1007/978-3-030-01941-9_3

index of fine-grained soils provide valuable information relative to potential foundation settlements and are viewed as a substitute for conducting consolidation tests. Further, it is often desirable to obtain approximate values using other soil indices which are more easily determined than standard consolidation tests. Approximate values are important in preliminary studies of settlement and provide some indication of order of magnitude of compression index when conducting consolidation tests. Therefore, it is imperative that the selected soil index property is easy to determine, don't require extensive time and/or high expense.

The majority of available empirical models relate the compression index to soil liquid limit, natural water content, and in-situ void ratio. Such approximations rely on the use of traditional regression analysis using consolidation test data and developing expressions relating the compression index to soil index properties. Some authors have developed multiple regression models that relate the compression index to soil liquid limit and void ratio (Al-Khafaji and Andersland 1992). Such models require more laboratory tests and expense than a single independent variable such as the liquid limit, LL. Most of the published models fail to consider data outliers and their impact on derived regression expressions. In some cases, a few data outliers can significantly alter the proposed model. Such data outliers may be associated with high liquid limits. More recently, Artificial Neural Networks (ANN) tools have been used to develop more reliable regression models. Such models often rely on limited data and ignore data outliers. Although ANN is a valuable tool, it is data driven and the quality and quantity of the data used impact derived models.

Empirical relationships relating compression index of fine-grained soils to their liquid limit have been suggested. Among the more widely used equations for estimating compression index are (Skempton 1944; Terzaghi and Peck 1948). Other less well-known equations include (Helenelund 1951; Nishida 1956; Koppula 1981). Besides statistical trends, these empirical equations do not seem to have a logical and/or theoretical basis to justify their development. For example, the Terzaghi regression equation was based on 36 consolidation test data performed on Chicago clay soils. Further, the Skempton approximation is only applicable to remolded clay. The focus of this paper is on newly published regression approximations relating fine-grained soil compression index to its liquid limit. The list of such published approximations continues to grow with additional statistical measures such as Robust Regression and standard errors. The weaknesses of the traditional least squares regression estimates are related to data outliers, choice of regression model used and error associated with data collected.

2 New Empirical Compression Index Approximations

An exhaustive study of more recently published compression index approximations of the compression index in terms of the liquid was conducted which produced a total 1906 data points. Further, the statistical accuracy of the models relating soil Compression Index, Cc to its Liquid Limit, LL were also examined. Published statistical measures included correlation coefficients and standard errors. A set of more recently proposed approximations for soil Compression Index in terms of its Liquid Limit are shown in Table 1.

Table 1. Recently proposed approximations for soil Compression Index in terms of its Liquid Limit.

No	Regression equation	R^2	Author
1	Cc = 0.009 (LL – 10)	0.68	Terzaghi and Peck (1948)
2	Cc = 0.008 (LL – 12)	0.78	Sridharan and Nagaraj (2000)
3	Cc = 0.0055 (LL – 1.836)	0.97	Vinod and Bindu (2010)
4	Cc = 0.01706 (LL – 1.30)	0.35	Widodo and Ibrahim (2012)
5	Cc = 0.0046 (LL – 1.39)	0.99	Laskar and Pal (2012)
6	Cc = 0.004 (LL – 7.5)	0.78	Ofosu (2013)
7	Cc = 0.0118 (LL – 20.7)	0.81	McCabe et al. (2014)
8	Cc = 0.0027 (LL + 73.85)	0.25	Dway and Thant (2014)
9	Cc = 0.002 (LL – 63.5)	0.91	Nesamatha and Arumairaj (2015)
10	Cc = 0.0067 (LL – 5.43)	0.94	Kumar et al. (2016)
11	Cc = 0.005 (LL + 6.4)	0.57	Jacob and Hari (2016)
12	Cc = 0.012 (LL – 8)	0.87	Kootahi and Moradi (2017)

Although Eqs. 1, 2 and 12 shown in Table 1 have similar slopes and intercepts, other equations vary widely depending on the data used and geographic location. An obvious question can be raised as to which one of these expressions is most reliable. Needless to say, there is no clear answer because these approximations are based on test data that vary in terms of tests reliability, number of data points used and range of liquid limit values involved. For example, using a few data points may produce a high correlation coefficient between soil compression index and its liquid limit but will also produce a large standard error. On the other hand, a large number of data points may produce a low correlation coefficient but a smaller standard error. Furthermore, the presence of data outliers could significantly impact the derived expression. In one case, the authors removed a single data point outlier which significantly reduced the correlation coefficient. The variability of the various empirical models listed in Table 1 are shown in Fig. 1.

It is interesting to note that Terzaghi's empirical equation represents somewhat of average for all the newly published approximations as shown by the dotted line in Fig. 1. The Terzaghi expression $C_c = 0.009(LL-10)$ was based on 36 data points from the Chicago area with a correlation coefficient of 0.68. Also, Eq. (9) gives negative compression indices for liquid limits below 63.5! Clearly, published empirical approximations of soil compression index in terms of its liquid limit are useful but require sound engineering judgment and experience. Further, combining consolidation data for highly compressible soils with those of mineral soils diminishes the applicability of some published models. Data outliers associated with highly compressible soils were common in many of the data sets examined in this article but were dealt with using Robust Bi-Square Regression.

N

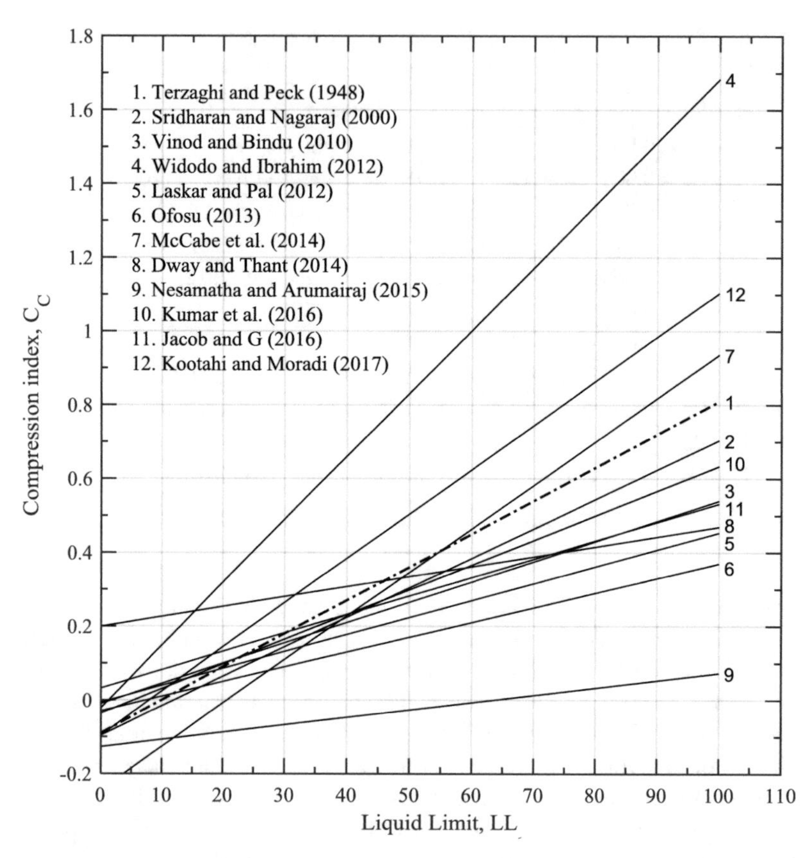

Fig. 1. Graphical representation of newly published regression equations relating soil compression index to its liquid limit.

It was shown that use of compression index in settlement calculation of organic soils is not justified (Al-Khafaji and Andersland 1981). However, for a majority of practical problems relating to mineral soils, consolidation pressure has a minor influence on compression index. Examination of published data and compression index equations lead to the conclusion that such relationships are only approximate. The overwhelming evidence is that clay structure, geological history, clay type, and presence of organic matter strongly influence soil compression index. New empirical expressions show that regression coefficients used in compression index equations are not unique.

3 Examination of Published Compression Index Data

In order to check the validity of the data sets published by various authors as listed in Table-1 was achieved using the Plasticity Chart. Unfortunately, the number of data points available was reduced from the original 1,906 data points to just 342 data points because many authors only provided liquid limits values but not the associated plastic limits. Therefore, imposing well-known liquid limits and plastic limit as reflected by the A-line and U-Line shown by Eqs. (1) and (2), the validity of published data was examined. That is,

$$A - \text{Line: } PI = 0.73\,(LL - 20) \quad (1)$$

$$U - \text{Line: } PI = 0.9\,(LL - 8) \quad (2)$$

The A-line is used to identify clay soils when the Plasticity Index, PI and the Liquid Limit, LL plot above the A-Line and Silty soils when PI and LL plots below the A-Line. The U-Line represents limiting values for LL and PI for any give soil in that no soil has ever been found to possess a liquid limit and plastic limit that plots above the U-Line. The CL-ML area has limits of plasticity index of 4 to 7 with an absolute minimum Liquid Limit of 16. The published data sets contained only 342 data points of Liquid Limit and Plasticity Index values. These data points were superimposed on the Plasticity Chart to test the validity of the combined data as shown in Fig. 2.

Careful consideration of Fig. 2 reveals that several data points plotted below the CL-ML area and at least one data point plotted above U-Line which is impossible. Also, several data points had liquid limit values of less than 16. These data outliers cannot just simply be removed from the data sets being considered because it will bring into question the reliability of the proposed regression models listed in Table 1. More importantly, such deviations of published data sets from well-established limits for the Liquid Limit are improper and should not have been permitted in determining the regression equation approximation of compression index in terms of its liquid limit.

Note that the data set plotted in Fig. 2 included liquid limit beyond 100 but were not shown to help focus on the range of liquid limit values most expected in engineering practice ($16 < LL < 100$). The collected data included Liquid limit and plasticity indexes beyond this range.

4 Validity of Empirical Compression Index Equations

Published equations for compression-index approximations are based in some cases on incorrect data that violate known limits for soil properties. In the case of soil liquid limit, many invalid data points were used that plotted above the U-Line and some below the CL-ML area as is shown in Fig. 2. However, the cumulative data set of 1,906 data points relating soil compression index to its liquid limit was by far the largest ever collected and required a closer look. Using The Bi-Square Robust regression on MATLAB, the entire data set was plotted and a linear model which

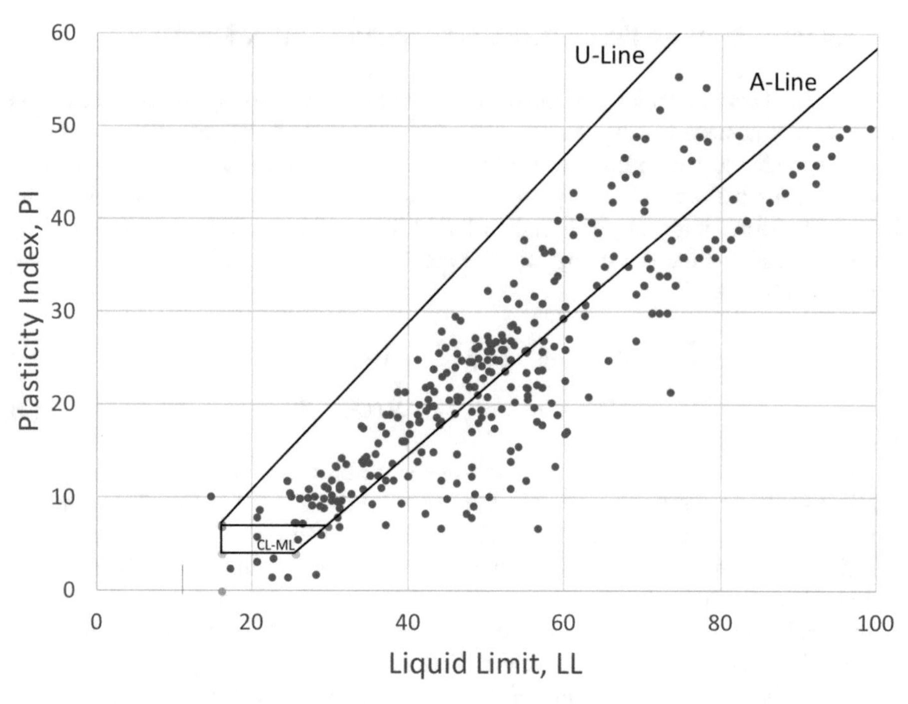

Fig. 2. The Plasticity Chart showing 342 published data points used in the development of in regression equations shown in Table 1.

relates compression index C_c to Liquid limit LL of the form given by Eq. (3) is assumed.

$$C_c = \alpha_c + \beta_c LL \tag{3}$$

Where α_L and β_L are the regression coefficients relating compression index to Liquid limit determined for a given data set. At first, traditional linear regression analysis was used and Robust regression was then performed using the combined data set 1906 data points as shown in Fig. 3.

Objectivity and an unbiased analysis require that one must not be selective in choosing data points used in regression analysis. For this reason, the full range of liquid limit values were used and corresponding regression equations calculated. The resulting regression coefficients (α_L and β_L) are shown below. Thus, using the entire data set of 1,906 data points with LL > 16 produced the following model:

$$C_c = -0.1096 + 0.01049\,LL \tag{4}$$

$R^2 = 0.7688$
SSE: 120.8
RMSE: 0.2519

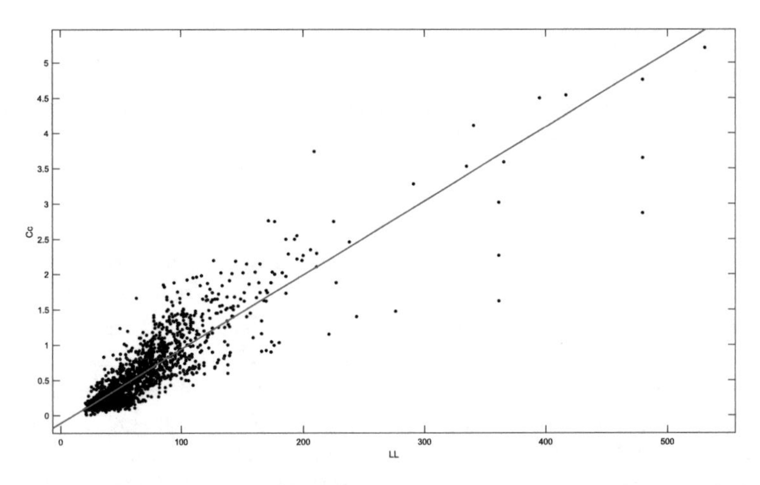

Fig. 3. Graphical illustration of the 1,906 data points relating the soil Compression Index, Cc to its Liquid Limit, LL by the various authors listed in Table 1.

Where
R^2 = The correlation between the response values and the predicted values.
SSE = Summed Square of Residuals and is usually labeled.
RMSE = Root Mean Squared Error or standard error.

Note that R^2 is also called the multiple correlation coefficient and the coefficient or multiple determination. That is, the higher the values indicate a strong correlation between the Compression Index and the Liquid Limit. Smaller RMSE values indicate that the model has a smaller random error component, and that the fit is more reliable. These values were associated with 95% confidence limit.

Consideration of the many data outliers shown in Fig. 3 mandated that a Robust regression analysis be used to minimize their effects on the derived regression equation. The Bi-Square Robust Regression found in MATLAB was used to produce a better correlation coefficient as show below.

$$C_c = -0.2006 + 0.01181\ LL \tag{5}$$

$R^2 = 0.8617$
SSE: 72.25
RMSE: 0.1948

Equation (5) produced less standard errors as reflected by the lower RMSE value and highlights the need to use Robust statistical analysis to develop better regression approximations. Examination of Eqs. (4) and (5) reveals that the Robust Bi-Square regression analysis produced more reliable estimates for the regression coefficients relating soil compression index to its liquid limit. This is readily apparent because of the higher R^2 value and the lower associated SSE and RMSE values obtained.

For All Soils with Liquid Limit: LL > 17

Unlike traditional regression techniques, Eq. (5) was developed by minimizing the impact of data outliers on the developed model. It can be rewritten in a simpler form for all 1906 soils samples used in the Bi-Square Regression analysis as follows:

$$C_c = -0.012\,(LL - 17) \tag{6}$$

Equation (6) is valid for soils with liquid limits of greater than 17. It relies on the combined data and not just some selected range of Liquid Limits that may be affected by local geology and other errors highlighted earlier. What is truly noteworthy is that Liquid limits < 17 will produce a negative compression index which is meaningless as it should. This is consistent with the lowest Liquid Limit of 16 expected for all soils. Finally, comparing Eq. (6) to the Kootahi and Moradi (2017) equation listed in Table 1 shows that their slopes are identical but not their intercepts. The difference is the intercepts may be explained by the fact that Eq. (6) is based on 1906 data points while Kootahi and Moradi (2017) was based on 500 data points. Clearly, their expression would yield a positive compression index for liquid limits of between 8 and 16 which is impossible.

For all practical purposes, the relationship between the compression index for fine-grained soils and the Liquid Limit is most often is limited to less than 100. It is well known that using the plasticity chart associated with the Unified Soil Classification System, a Liquid limit of greater than 100 is associated with highly compressible soils and organic soils. Most mineral soils are associated with liquid limits between 16 and

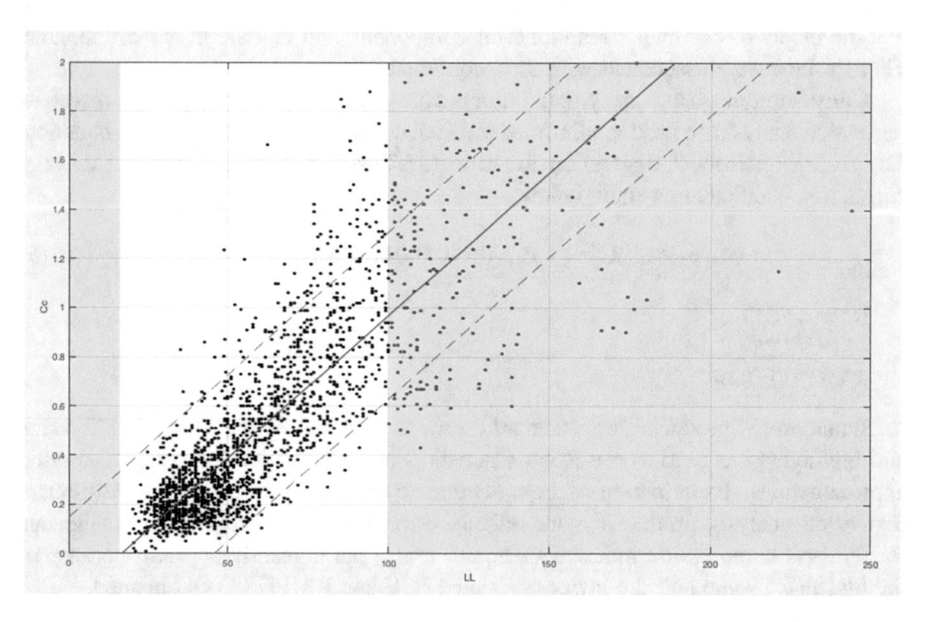

Fig. 4. Graphical illustration of the 1,906 data points relating the soil Compression Index, Cc to its Liquid Limit, LL by the various authors listed in Table 1.

100. Using MATLAB to perform the Robust Bi-Square Regression for soils with LL < 100 is shown graphically in Fig. 4 for the data used in the analysis.

The shaded area indicates the data points excluded from the Robust Regression analysis. The scatter in the data is clear for entire range of the liquid limits values used. This is true for both mineral soils with low plasticity and highly compressible soils. However, Fig. 4 shows the data scatter used in the Bi-Square analysis for Liquid limit between 16 and 100. Furthermore, the dashed lines represent the 95% confidence limits for the Robust Bi-Square regression coefficients given below.

$$C_c = -0.1927 + 0.01161\,LL \tag{7}$$

$R^2 = 0.7066$
SSE: 49.9
RMSE: 0.1728

Clearly, the R^2 value is lower than that obtained for the entire data. Also, the standard errors in Eq. (7) is lower than that obtained for the entire data. The advantage of Eq. (7) over traditional linear regression models found in literature is that it is based on the Robust Bi-Square Regression method which diminishes the impact of data outliers. The models shown in Table 1 are based on traditional regression techniques and ignore the impact of data outliers on derived regression coefficients and corresponding models.

For Soils With Liquid Limit: 16 < LL < 100

Typically, most mineral soils are associated with liquid limits of less than 100. Thus, Eq. (7) can be expressed for soils with Limited Liquid limit values between 16 and 100 as follows:

$$C_c = 0.0116\,(LL - 16.6) \tag{8}$$

What is interesting about Eqs. (8) and (6) is that they have remarkably similar slopes and intercepts. The implication is that the derived regression equations for all soils or soils with liquid limits between 16 and 100 are essentially the same! In fact, if we round off the regression coefficients, the equations are identical. However, the correlation coefficient for Eq. (8) of 0.7066 is lower than the 0.86 obtained for Eq. (6). Comparison of Eqs. (6) and (8) with those listed in Table 1 reveal that they both have the same slope as that reported by Kootahi and Moradi (2017). It is truly remarkable that the Terzaghi equation shown as a dotted line in Fig. 1 plots as an average between all of the newly derived equations for estimating the compression index in terms of the liquid limit for fine-grained soils. Yet the Terzaghi Equation was based on a few data points!

5 Conclusions

Published equations for compression-index approximations in terms of soil liquid limit show considerable inconsistencies. The differences are directly attributable to the nature of the data used in the development of empirical relationships and data outliers.

Often, information pertaining to soil type, regression coefficients and standard errors are missing. A five-year comprehensive literature search produced 1906 published data points by various authors. Examination of published various data sets indicates that removing only few data points from sets used in traditional regression analysis dramatically impacted published models. Such models fail to take into account data outliers. Further, some published empirical models included erroneous data that violate known limits of the liquid limit on the Plasticity Chart.

More recently, Artificial Neural Networks (ANN) have been employed to develop regression models relating soil compression index to its liquid limit. However, such models often involve limited number of data points and ignore data outliers. The authors employed the Robust Bi-Square regression technique in MATLAB to reduce the impact of data outliers and reduce standard errors. Several new empirical models were developed for approximating soil compression index in terms of its liquid limit with high correlation coefficient and low standard error. The proposed model given by Eqs. (6) is based on a combined 1,906 data points using Robust Regression that diminishes the effects of data outliers. The proposed model $C_c = 0.012(LL\text{-}17)$ is applicable to all soils with Liquid Limits of greater than 17. A second model given by Eq. (8) was developed using Robust Bi-Square regression and data sets with liquid limit between 16 and 100. This model $C_c = 0.0116(LL\text{-}16.6)$ is valid for liquid limits larger than 17. The regression coefficients associated with Eqs. (6) and (8) are very similar. Comparison of Eqs. (6) and (8) with those listed in Table 1 reveal that they both have the same slope as that reported by Kootahi and Moradi (2017). It is truly remarkable that Terzaghi's model of $C_c = 0.009(LL\text{-}10)$ shown as a dotted line in Fig. 1 represents an average for all models shown in Table 1 for estimating soil compression index in terms of its liquid limit. Yet, the Terzaghi model was based on 36 data points! Unlike traditional regression models, the new proposed regression models reduce the impact of data outliers on the regression coefficients and produce lower standard errors. This is important in that many other published models yield positive compression indices for liquid limits between 0 and 17 and involve higher standard errors due to the presence of data outliers.

Acknowledgments. We thank Caterpillar Inc. for providing funding to complete this research project through the Caterpillar Fellowship program established at Bradley University.

References

Al-Khafaji, A.W., Andersland, O.B.: Compressibility and strength of decomposing Fiber/Kaolinite. Geotechnique **31**(4), 497–508 (1981)

Al-Khafaji, A.W., Andersland, O.B.: Equations for compression index approximation. J. Geotech. Eng. Div. A.S.C.E., **118**(GT1) (1992)

Dway, S.M.M., Thant, D.A.A.: Soil compression index prediction model for clayey soils. Int. J. Sci. Eng. Tech. Res. **3**(11), 2458–2462 (2014)

Helenelund, K.V.: On consolidation and settlement of loaded soil-layers. Thesis presented to Finland Technical Institute, at Helsinki, Finland, in partial fulfillment of the requirement for degree of doctor of Philosophy (1951)

Kootahi, K., Moradi, G.: Evaluation of compression index of marine fine-grained soils by the use of index tests. Mar. Georesour. Geotechnol. **35**(4), 548–570 (2017)

Koppula, S.D.: Statistical estimation of compression index. Geotech. Test. J. GTJODJ **4**(2), (1981)

Kumar, R., et al.: Prediction of compression index (Cc) of fine grained remolded soils from basic soil properties. Int. J. Appl. Eng. Res. **11**(1), 592–598 (2016)

Jacob, K., Hari, G.: Study on the relationship of compression index from water content, atterberg limits and field density for Kuttanad clay. Int. J. Innov. Res. Tech. **3**(4), 33–38 (2016)

Laskar, A., Pal, S.K.: Geotechnical characteristics of two different soils and their mixture and relationships between parameters. Electron. J. Geotech. Eng **17**, 2821–2832 (2012)

McCabe, B.A., et al.: Empirical correlations for the compression index of Irish soft soils. Proc. Inst. Civ. Eng. Geotech. Eng. **167**(6), 510–517 (2014)

Nesamatha, R., Arumairaj, P.D.: Numerical modeling for prediction of compression index from soil index properties. Electron. J. Geotech. Eng. **20**, 4369–4378 (2015)

Nishida, Y.: A brief note on compression index of soil. J. Soil Mech. Found. Eng. Div. ASCE, **82** (SM3). In: Proceedings Paper 1027, 1027-1–1027-14 (1956)

Ofosu, B.: Empirical model for estimating compression index from physical properties of weathered birimian phyllites. Electron. J. Geotech. Eng. **18**, 6135–6144 (2013)

Skempton, A.W.: Notes on the compressibility of clays. Q. J. Geol. Soc. Lond. **100**, 119–135 (1944)

Sridharan, A., Nagaraj, H.B.: Compressibility behavior of remoulded, fine-grained soils and correlation with index properties. Can. Geotech. J. **37**, 712–722 (2000)

Terzaghi, K., Peck, R.: Soil Mechanics in Engineering Practice. Wiley, New York, N.Y. (1948)

Widodo, S., Ibrahim, A.: Estimation of primary compression index (Cc) using physical properties of pontianak soft clay. Int. J. Eng. Res. Appl. **2**(5), 2232–2236 (2012)

Validation of Compression Index Approximations Using Soil Void Ratio

Amir Al-Khafaji[(✉)], Abbey Buehler, and Ethan Druszkowski

Department of Civil Engineering and Construction, Bradley University, Illinois, USA
amir@fsmail.bradley.edu

Abstract. Published equations for approximations of soil compression index in terms of in situ void ratio are varied and confusing. Many published expressions yield a positive compression index even for soils with zero void ratios. Furthermore, published expressions are normally based on a high correlation coefficient and limited test data. Significant measures such as standard errors are often omitted. Some published equations are limited to fine-grained or organic soils. Most available data exhibit a behavior in which significant scatter is observed for small values of void ratio. High void ratio values associated with organic soils produce high compression indices. The use of compression index in settlement calculation of organic soils deposits and especially peat is not justified. Over the past five years, significant amount of data was collected from published articles in reputable journals and conferences. A total of 1722 data points relating the compression index to soil void ratio were analyzed. Using MATLAB, the authors developed new regression models that relate the compression index of normally consolidated soils to its void ratio and including careful examination of data outliers. Using the methods of Robust Regression, the proposed regression models minimized the effects of data outliers and produced higher correlation coefficients with reduced associated standard errors. Although several artificial neural networks (ANN) regression models are available, such models were based on limited data without dealing with data outliers. The combined data set used is then largest ever assembled and offers clear insights into the behavior of mineral and organic soils. The proposed regression models differ from published empirical approximations by considering the impact of data outliers.

Keywords: Compression index · Void ratio · Regression · Empirical Data · ANN · Validity · MATLAB

1 Introduction

The consolidation settlement of foundations placed on normally consolidated fine-grained soil deposits requires knowledge of soil compression index, C_c and the in-situ void ratio, e_o. Both of these soil parameters involve standardized consolidation tests which are both time consuming and expensive. The number of tests required is dependent on the variability of the soil profile and stress increments involved. The development of empirical equations for estimating the compression index of fine-

S. Hemeda and M. Bouassida (Eds.): GeoMEast 2018, SUCI, pp. 42–52, 2019.
https://doi.org/10.1007/978-3-030-01941-9_4

grained soils in terms of its void ratio provide valuable information relative to potential foundation settlements. Further, it is often desirable to obtain approximate values using other soil indices which are more easily determined than standard consolidation tests. The soil void ratio can be calculated using basic laboratory tests. Therefore, it is possible to estimate the soil compression index without the extensive time and/or high expense associated with consolidation tests. The majority of available empirical formulas relate the compression index to soil liquid limit, natural water content, and in situ void ratio. Such approximations rely on the use of regression analysis using consolidation test data and developing regression equations to relate the compression index to soil index properties. Some authors have developed multiple regression models that relate the compression index to soil liquid limit and void ratio (Al-Khafaji, et al. 1992). Such models require more laboratory tests and expense than a single independent variable such as the void ratio.

Empirical relationships relating compression index of fine-grained soils to their liquid limit have been suggested. Linear and nonlinear empirical equations have been published that relate compression index to soil void ratio. Some published expressions are applicable to all soils while others are limited to specific soil types and/or geography (Peck and Reed 1954). Authors typically provide the correlation coefficient (R^2) as a lone measure to justify the derived expressions applicability to a wide range of soils. The number data points used and the standard error are often omitted which raises serious problems in that a few data points may produce extremely high R^2 value but lack reliability. Additionally, the methods used in data collection and data analysis were varied and introduced uncertainty relative to the accuracy and dependability of published equations for compression index approximation. The recent publication of a substantial number of consolidation data and empirical equations warrants a thorough examination of published methods proposed for approximating the compression index (C_c) of soil in terms of its void ratio (e_o). The authors collected over 1722 data points involving the compression index and void ratio. This data set was used to develop new expressions to approximate the Compression Index in terms of its void ratio. Furthermore, using MATLAB, the authors were able to utilize Robust Regression analysis to minimize the effect of data outliers. Data outliers are ignored when using traditional linear regression associated with published models relating C_c to e_o.

Familiar equations to estimate C_c in terms of the void ratio include Nishida (1956), Hough (1957), and Bowles (1989). Examination of 3-D models indicate that consolidation pressures and organic content significantly influence the Cc value. Al-Khafaji and Andersland (1981) demonstrated that the use of C_c in settlement analysis for organic soils is not appropriate. The normal approach of combining mineral and organic soils data without examining data outliers is not appropriate.

2 Published Empirical Compression Index Approximations

In practice, engineers often rely on empirical approximations to estimate the C_c in terms of soil index properties because they are less expensive than consolidation tests and provide meaningful initial estimates for settlement of foundations. Such estimates are only valid when dealing with normally consolidated fine-grained soils unless the pre-consolidation pressure and recompression index are known. The focus of this paper is on published linear empirical equations used to estimate C_c in terms of its in-situ void ratio. An extensive literature search was undertaken to identify best known and newest approximations of the compression index of normally consolidated clay as shown in Table 1.

Table 1. Empirical equations for compression index approximation using void ratio.

No	Regression equation	R^2	Author
1	$C_c = 0.54\ (e_o - 0.35)$	–	Nishida (1956)
2	$C_c = 0.35\ (e_o - 0.5)$	–	Hough (1957)
3	$C_c = 0.43\ (e_o - 0.25)$	–	Cozzolino (1961)
4	$C_c = 0.75\ (e_o - 0.50)$	–	Sower and Sower (1970)
5	$C_c = 0.40\ (e_o - 0.25)$	–	Azzouz et al. (1976)
6	$C_c = 0.156\ e_o + 0.0107$	–	Bowles (1989)
7	$Cc = 1.02 - 0.95\ e_o$	–	Gunduz and Arman (2007)
8	$Cc = 0.2875\ (e_o - 0.5082)$	0.90	Vinod and Bindu (2010)
9	$Cc = 0.5217(e_o - 0.20)$	0.43	Widodo and Ibrahim (2012)
10	$Cc = 0.3608\ e_o - 0.0713$	0.96	Kalantary and Kordnaeij (2012)
11	$Cc = 0.196\ e_o + 0.207$	0.56	Dway and Thant (2014)
12	$Cc = 0.510\ (e_o - 0.33)$	0.92	Kootahi and Moradi (2017)

Note that some of the earlier publications reported no values for the correlations coefficient, R^2. A closer look at the published regression equations listed in Table 1 indicate some similarity between the regression coefficients and some drastic differences. For example, Eqs. 3 and 5 essentially have the same regression coefficients. Also, Eq. 7 is the only expression that proposes a lower value for the compression index with increasing void ratio! Furthermore, several equations yield a negative compression index for fine-grained soils with void ratios of less than 0.5. This is not reasonable for some clays knowing that the minimum void ratio may be as low as 0.26 (Terzaghi et al. 1996).

An obvious question can be raised as to which one of these expressions is reliable in approximating the compression index in terms of void ratio for clay. Obviously, there is no clear answer because these approximations are based on test data that vary in terms of local geology, tests reliability, number of data points used, human error, and range of void ratio included. For example, using a few data points may produce a high correlation coefficient between the compression index and the void ratio but will also produce a large standard error. On the other hand, a large number of data points may produce a low correlation coefficient but a smaller standard error. The empirical equations listed in Table 1 are illustrated graphically in Fig. 1.

The published equations presented in Table 1 and shown in Fig. 1. are all linear and developed using regression analysis. It is expected, that the compression index increases with increasing void ratio for all expressions shown in Fig. 1 Except Eq. (7) reported by Gunduz and Arman (2007). Thus, this expression makes no practical sense

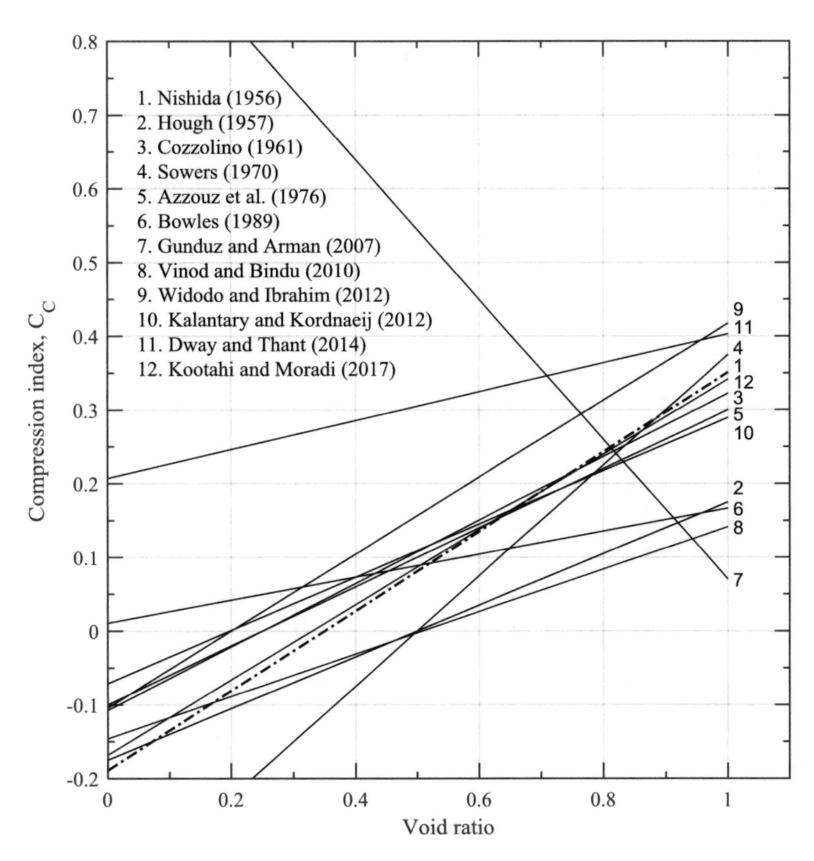

Fig. 1. Graphical representation of newly published regression equations relating soil compression index to its liquid limit.

and should be ignored. Note that the Nashida equation is shown by a dotted line for reasons to be explained later. Hough (1957) concluded that important differences exist between organic and mineral clay soils. Azzouz et al. (1976) proposed an expression for all natural soils. Lambe and Whitman (1969) showed that empirical expressions were not reliable, based on a linear correlation between the ratio $C_c/(1 + e_0)$ versus natural water content. Some authors have proposed expressions for specific geographic areas (Cozzolino 1961). In all cases, the slopes and/or intercepts vary significantly depending on the author and the test date used.

It was shown that use of compression index in settlement calculation of organic soils is not justified (Al-Khafaji and Andersland 1981). Examination of published data and compression index equations lead to the conclusion that any relationship between compression index and soil index properties is only approximate. The overwhelming evidence is that clay structure, geological history, clay type, and presence of organic matter strongly influence soil compression index. New empirical expressions show that regression coefficients used in compression index approximation are not unique.

3 Examination of Published Compression Index Data

The validity of the data and expressions published by various authors listed in Table-1 was achieved by plotting the void ratio versus the liquid limit for the collected data. The number of data points available for analysis by the authors was 1,722 data points as shown in Fig. 2.

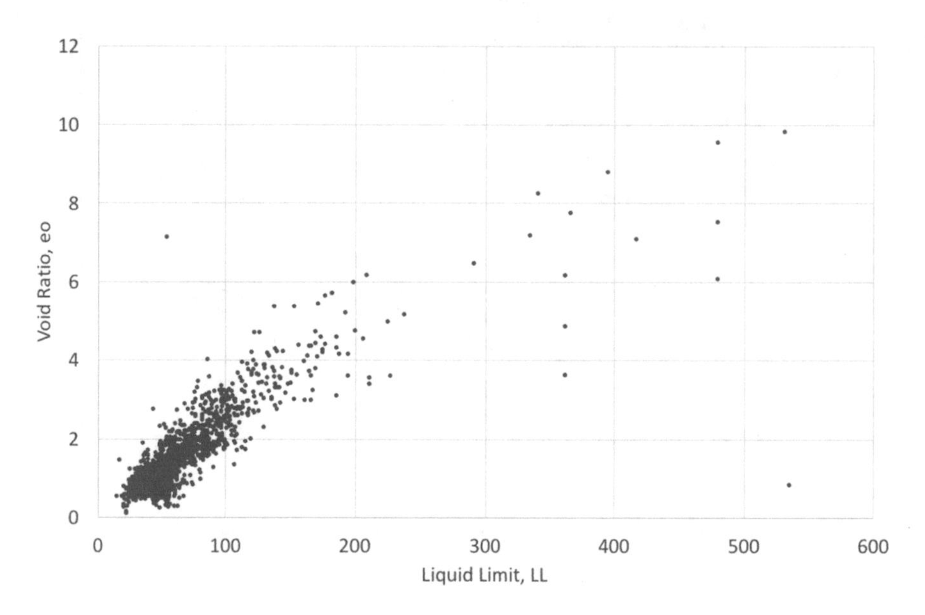

Fig. 2. Graphical illustration showing 1,722 published data points relating the void ratio to the liquid limit for equations shown in Table 1.

Careful examination of Fig. 2 reveals that there were several data point outliers. For example, one data point would give a compression index of more than 7 for a liquid limit of less than 70! Additionally, another data point with a liquid limit of more than 500 is associated with a compression index of less than 1. These data points and others shown in Fig. 2 indicate that some of the published data and compression index approximations are not reliable. In fact, the authors have examined some of these data sets for some authors and discovered that removing even one data point produced a much lower correlation coefficient than those reported in the literature. Also, several published data points for C_c and e_o are associated with liquid limit values of less than 16 or above the U-line found in the Plasticity Chart. In such cases, the corresponding void ratio values reported by authors cannot be trusted. Such data outliers cannot just simply be removed from the data sets being considered because it will bring into question the reliability and bias of the proposed regression models published as listed in Table 1.

4 Robust Regression to Reduce Outlier Effects

The published regression models described in Table 1 are based on certain assumptions, such as a normal distribution of errors of the published data. If the distribution of errors is prone to data outliers, model assumptions are invalidated, and published statistical models become unreliable. The robust fitting method is less sensitive than ordinary least squares to data outliers. Although several artificial neural network (ANN) regression models are available, such models were based on limited data without dealing with data outliers.

One remedy to deal with outliers is to remove data outliers from the least-squares fit. This is not acceptable because it introduces bias into modeling and permit authors to simply throw away data that don't meet their expectations. Another approach, termed robust regression, is to use a fitting criterion that is not as vulnerable as least squares to unusual data outliers. The most common general method of robust regression is M-estimation. This class of estimators can be regarded as a generalization of maximum-likelihood estimation, hence the term "M"-estimation. MATLAB offers Robust regression based on the method developed by Huber and Tukey. The authors used the Tukey Bi-Square technique to reduce the impact of data outliers.

Robust regression works by assigning a weight to each data point. Weighting is done automatically and iteratively. In the first iteration, each point is assigned equal weight and model coefficients are estimated using ordinary least squares. At subsequent iterations, weights are recomputed so that points farther from the proposed model predictions in the previous iteration are given lower weight. Model coefficients are then recomputed using weighted least squares. The process continues until the values of the coefficient estimates converge within a specified tolerance specified by the author.

5 Validity of Empirical Compression Index Equations

Published equations for compression-index approximations are based in some cases on incorrect data that violate know limits for soil index properties. In the case of soil liquid limit, many invalid data points were used that plotted above the A-Line and some below the CL-ML area on the plasticity chart. In such cases, the corresponding void ratio data points associated with the liquid limit produced data outliers that impacts the derived expressions as shown in Fig. 2. Using MATLAB, the data set was plotted and a linear model which relates the soil compression index C_c to it void ratio, e_o as given by Eq. (3).

$$C_c = \alpha_e + \beta_e e_o \tag{1}$$

Where α_e and β_e are the regression coefficients relating C_c to e_o for a given range of void ratios for a given data set. Regression analysis was then performed using the combined data set 1,722 data points as illustrated graphically in Fig. 3.

Clearly, Fig. 3 reveals that a linear correlation between the compression index and its void ratio is feasible. Objectivity and an unbiased analysis require that one must not be selective in choosing data points used in regression analysis. For this reason, the full

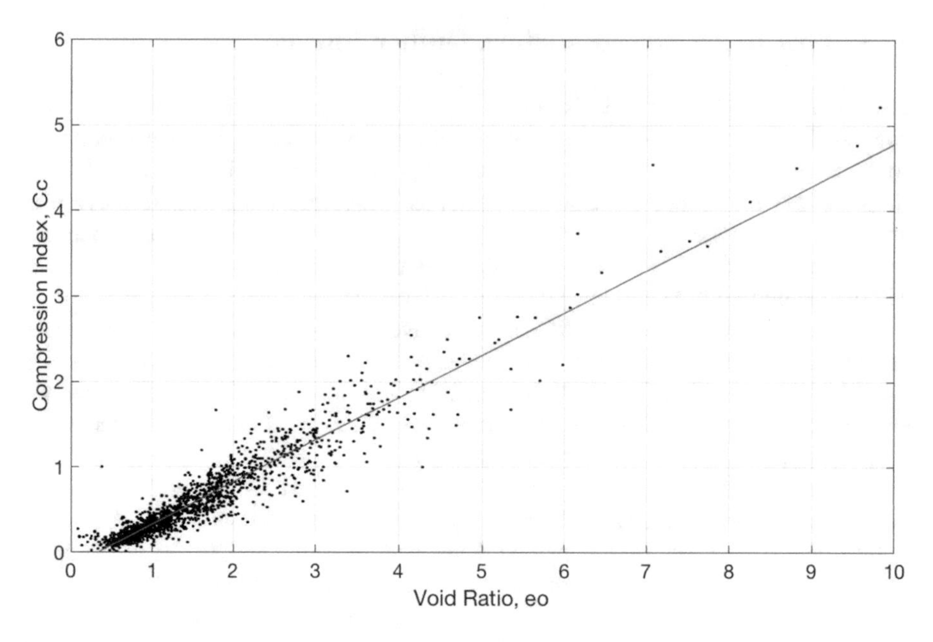

Fig. 3. Graphical illustration of the data points relating the soil Compression Index, Cc to its Void Ratio, e_o by the various authors listed in Table 1.

range of liquid limit values were used and corresponding regression equations calculated. The resulting regression coefficients α_e and β_e. Thus, using the entire set of 1,722 data points with LL > 16 produced the following regression equation:

$$Cc = -0.1609 + 0.4942\, e_o \tag{2}$$

$R^2 = 0.9077$
SSE: 46.56
RMSE: 0.1646

Where

R^2 = the correlation between the response values and the predicted values.
SSE = Summed Square of Residuals.
RMSE = Root Mean Squared Error and it is also referred to as standard error.

These statistical measures were associated with 95% confidence limit. Consideration of the many data points outliers mandated the use of Robust regression analysis to minimize their effects on the derived regression model. The Bi-Square Robust Regression found in MATLAB was used to produce a better correlation coefficient as show below.

$$Cc = -0.1687 + 0.5005\, e_o \tag{3}$$

$R^2 = 0.9496$
SSE: 25.42
RMSE: 0.1216

Note that R^2 value has increased from 0.91 to 0.95 and the corresponding errors is reduced from 0.16 to 0.12. That is, the higher the value indicates a stronger correlation between the Compression Index, C_c and the Void Ratio, e_o. A smaller RMSE value indicates that the proposed model is more reliable than that developed using the standard regression procedures.

Examination of Eqs. (2) and (3) shows that the Robust Bi-Square analysis produced more reliable estimates for the regression models. This is readily apparent because of the higher R^2 value and the lower associated SSE and RMSE values obtained using Robust regression reduced the impact of data outliers. Thus, the authors suggest using a simplified expression for the approximation by rewriting Eq. (3) as follows:

Valid for Void Ratio: $e_o > 0.35$

$$C_c = 0.5\,(e_o - 0.34) \tag{4}$$

Equation (4) is based on 1,722 data points and is applicable for all soils. The regression coefficients shown in Eq. (4) are very close to those reported by Nashida (1956) and Kootahi and Moradi (2017). The Nashida equation was shown by a dotted line in Fig. 1. It is interesting that the Nashida expression published in 1956 appears to be more accurate than other approximations reported by many other authors and provide an average between all other expressions shown in Fig. 1.

Equation (4) is validity is supported by the large number of data points used, applying Robust Bi-Square regression and assuming a high confidence limit of 95%. The model developed by Kootahi and Moradi (2017) was based on more than 500 data points and provide further support for the model proposed by the authors. The proposed model given by Eq. (4) significantly reduces the impact of data outliers on the derived regression coefficients. Traditional linear regression analysis used by other others failed to address the problem of outliers. The proposed model is consistent with known minimum values of the void ratio expected for fine-grained soils. It will yield positive compression index values for soils with void ratio of greater that 0.34.

For all practical purposes, the relationship between fine-grained soils and the void ratio is most often is limited to void ratios of less than 2.5. Furthermore, this conclusion is based on Fig. 2 where a liquid limit of 100 will yield approximately a void ration of 2.5. It is well known that using the plasticity chart associated with the Unified Soil Classification System, a Liquid limit of greater than 100 is associated with fine-grained soils is considered highly compressible. Therefore, using MATLAB to perform Robust Bi-Square Regression on soils with void ratios of less 2.0 and excluding higher void ratio values produced the following regression equation:

$$C_c = -0.1378 + 0.4753\, e_o \tag{5}$$

$R^2 = 0.7942$
SSE: 24.52
RMSE: 0.1288

Note that the coefficient of correlation has dropped from 0.95 to 0.79 because of the reduced data set and data outliers. This approximation would more accurately reflect the expected behavior of most fine-grained soils encountered in practice. Thus Eq. 5 can be rewritten more simply as

Valid for Void Ratio: $0.30 < e_o < 2.50$

$$C_c = 0.475(e_o - 0.29)e_o \tag{6}$$

It should be noted that the compression index for fine-grained soils is positive only when the void ratio is greater than 0.29. This value compares closely with the minimum void ratio of 0.25 expected for fine-grained soils! It is recommended that Eq. 4 be used for all soils with any void ratio and that Eq. (6) be used only for soils with void ratios of less than 2.5 and more than 0.29.

6 Conclusions

Published empirical expressions used to estimate compression index of soils in terms of void ratio have been developed using linear regression analysis of laboratory data. The number of data points used and standard errors are often excluded which makes it difficult to assess their validity. Furthermore, the variability of soil parameters, soil types, and machine and operator errors makes it impossible to suggest a unified approach to compression index estimation in terms of the void ratio. In some cases, organic soils test data is included in the derivation of regression equations. However, organic soils properties change under constant effective consolidation pressure (Al-Khafaji and Andersland 1981) and should not have been used. Empirical expressions relating C_c to the void ratio should be limited to mineral soils.

Although several artificial neural network (ANN) regression models are available, such models were based on limited data without dealing with data outliers. New regression models are proposed for estimating the compression index in terms of soil void ratio using Robust Regression analysis that reduce the impact of data outliers. The general model is applicable to all soils as shown by Eq. 4 with void ratios of greater than 16 as dictated by the plasticity chart. A second model is limited to void ratios between 0.30 and 2.5 which closely corresponds to expected void ratio values for mineral soils. The validity of the proposed models is based on 1,722 data points gathered through an extensive literature search, use of Robust regression, careful examination of data input and comparing the results with other published models.

Equations (4) and (6) provide new approximations for the compression index and are based on the largest data set ever assembled and using Robust Regression. The proposed expressions demonstrate that several published empirical equations are not reliable. Although some published empirical equations are limited to mineral soils, others are purported to apply to all soils including organic soils. Use of these formulas is often legitimized based on the R^2 value without proper examination of the associated standard errors and number of data points used. Consideration of a number of widely-known empirical compression equations revealed significant variations exist between the slopes and intercepts of published expression due to geographic location, number of data points used, and data point outliers. Examination of data scatter reveals that high values of e_o are generally associated with organic soils and should not be included with data for mineral soils. The inclusion of such data points in derivations of empirical expressions alters the validity of these equations when dealing with mineral soils. Using MATLAB, the authors were able to minimize the effects of data outlier using Robust Bi-Square regression analysis.

Acknowledgments. We thank Caterpillar Inc. for providing funding to complete this research project through the Caterpillar Fellowship program established at Bradley University.

References

Al-Khafaji, A.W., Andersland, O.: Compressibility and strength of decomposing fibre-clay soils. Geotechnique **31**(4), 497–508 (1981)

Azzouz, A.S., Krizek, R.J., Corotis, R.B.: Regression analysis of soil compressibility. Soils Found. Jpn. Soc. Soil Mech. Found. Eng. **16**(2) (1976)

Bowles, J.E.: Physical and Geotechnical Properties of Soils. McGraw-Hill Book Co. Inc, New York, N.Y. (1989)

Cozzolino V.M.: Statistical forecasting of compression index. In: Proceedings of the 5th International Conference on Soil Mechanics and Foundation Engineering, Paris, vol. 1, pp 51–53 (1961)

Dway, S.M.M., Thant, D.A.A.: Soil compression index prediction model for clayey soils. Int. J. Sci. Eng. Tech. Res. **3**(11), 2458–2462 (2014)

Gunduz, Z., Arman, H.: Possible relationships between compression and recompression indices of a low-plasticity clayey soil. Arab. J. Sci. Eng. 32(**2B**), 179–190 (2007)

Hough, B.K.: Basic Soils Engineering, 1st edn. The Ronald Press Company, New York (1957)

Kalantary, F., Kordnaeij, A.: Prediction of compression index using artificial neural network. Sci. Res. Essays **7**(31), 2835–2848 (2012)

Kootahi, K., Moradi, G.: Evaluation of compression index of marine fine-grained soils by the use of index tests. Mar. Georesources Geotechnol. **35**(4), 548–570 (2017)

Lambe, T.W., Whitman, R.V.: Soil Mechanics. Wiley, New York, NY (1969)

Nishida, Y.: A brief note on compression index of soil. J. Soil Mech. Found. Eng. ASCE **82**(3), 1–14 (1956)

Peck, R.B., Reed, W.C.: Engineering Properties of Chicago Subsoils. Bulletin 423, Engineering Experiment Station, University of Illinois, Urbana, IL (1954)

Vinod, P., Bindu, J.: Compression index of highly plastic clays – an empirical correlation. Indian Geotech. J. **40**(3), 174–180 (2010)

Widodo, S., Ibrahim, A.: Estimation of primary compression index (Cc) using physical properties of pontianak soft clay. Int. J. Eng. Res. Appl. **2**(5), 2232–2236 (2012)

Terzaghi, K., Peck, R., Mesri, G.: Soil Mechanics in Engineering Practice. Wiley, New York (1996)

Sower, G.B., Sower, G.F.: Introductory Soil Mechanics and Foundations, 3rd Revised edn. Collier Macmillan Ltd., New York

Innovative Model for Settlement Calculations in Organic Soils

Amir Al-Khafaji(✉)

Department of Civil Engineering and Construction, Bradley University, Illinois, USA
amir@fsmail.bradley.edu

Abstract. The settlement of organic soils is dependent on the relative proportions of organic and mineral fractions and the degree of organic solids decomposition. The characteristics of the organic and mineral fractions significantly impact organic soils behavior. Experimental compression data for peat moss and mixtures of organic fibers with kaolinite or fine sand show a linear relationship between the organic fraction (by weight) and void ratio for constant stress levels. Over the range of possible organic content, this relationship is nonlinear for mixtures of organic fibers with montmorillonite. Decomposition in fibre-clay soils reduces volume of organic solids and produces changes in compressibility parameters. Use of the ignition test for measurement of the organic fraction permitted changes in the degree of decomposition to be monitored concurrently with changes in compressibility and vane shear strength. Void ratio parameters, based on experimental data, permitted calculation of equilibrium void ratios in terms of the organic fraction and pressure. An innovative model is proposed which permits calculation of equilibrium void ratios directly for any combinations of stress level and organic content. The proposed model eliminates the need for using compression index in settlement calculations associated with organic soils. Decomposition of organic solids at a constant over-burden pressure contributes additional volume change. Volume change caused by the loss of organic solids for soils mixtures involving organic fibers with sand, kaolinite, or montmorillonite can be readily predicted. New compression parameters for several soil mixtures revealed that organic soil behavior is complex and is dependent on organic fraction and type of mineral present.

Keywords: Settlement · Kaolinite · Montmorillonite · Decomposition Fibre-clay · Soils · Solids · Innovative · Model · Void ratio · Fraction

1 Introduction

The calculation of settlement of foundation placed on organic soils is a complex process. The traditionally methods used for computing settlement of foundations placed on mineral soils is still being used. Such approach is incorrect because of potential decomposition of the organic fraction and potential settlement without any changes in effective stress. Furthermore, organic soils are highly compressible depending on the organic fraction present in the soil profile and the potential decomposition of the

S. Hemeda and M. Bouassida (Eds.): GeoMEast 2018, SUCI, pp. 53–66, 2019.
https://doi.org/10.1007/978-3-030-01941-9_5

organic fraction associated with a given profile. The use of compression index in settlement calculations of organic soils is invalid because in soe cases, computed settlement values may exceed initial soil layer thickness!

Problems associated with decomposition in organic soils are often avoided by engineers because of a lack of information on the effects of decomposition on engineering properties. In many cases, decomposition rates are slow and the effects can be ignored. Additional settlement due to loss of organic solids by decomposition may go unnoticed. In cold regions where decomposition rates are small and humid areas were large amount of organic soil are present, there is a growing interest in organic soil behavior. Table 1 shows Peatlands of the Earth (Mersi 2007).

Table 1. Peatlands of the earth

Country	Peatlands, km^2	% of land area
Canada	1,500,000	18
U.S.S.R. the former	1,500,000	
United States	600,000	10
Indonesia	170,000	14
Finland	100,000	34
Sweden	70,000	20
China	42,000	
Norway	30,000	10
Malaysia	25,000	
Germany	16,000	
Brazil	15,000	
Ireland	14,000	17
Uganda	14,000	
Poland	13,000	
Falklands	12,000	
Chile	11,000	
Zambia	11,000	
26 other countries	220–10,000	
Scotland		10
15 other countries		1–9

Typically, Peat is composed of decomposed organic materials in excess of 30% by weight. Decomposition of organic soil fraction is dependent on the availability of nutrients, moisture and suitable temperatures. In field deposits, aerobic decomposition generally develops above the water table where sufficient nutrients and oxygen are available. Anaerobic decomposition may occur above or below the water table where there is an absence of oxygen, nutrients are present, pH is about 5.8–7.4 and anaerobes (bacteria) are present. The use of kaolinite and montmorillonite clay with pulp fibres in the preparation of fibre-clay soils has permitted precise control over organic and mineral fractions, fibre type and size. Nutrients added to the fibre-clay soils, in proportions similar to those found in an average bacterial cell (McKinney 1962), and storage of anaerobic samples at a temperature close to 35 °C between physical tests helped to accelerate the decomposition process. Decomposition measurement, using the ignition test is reviewed and a relationship is presented, which permits computation of the degree of decomposition. The effects of organic content and decomposition on soil compressibility are described. The conclusions summarize the effects of organic material and decomposition on the behaviour of fibre-clay soils

2 Experimental Work

The experimental work involved the preparation of fibre-clay soil samples followed by consolidation tests. Two different pieces of equipment were used to deal with low pressures and high-pressure load increments. A brief description of the soil materials, sample preparation, and consolidation tests and addressed next

2.1 Materials Studied

Using Pulp fibres which included a range of sizes with a weighted average length of 1·6 mm and typical diameters of about 20 μm were observed in the electron microscope. Surface area measurements, using the water vapour absorption method (Perkin-Elmer Corporation 1961), gave values close to 133 m^2/g of dry fibre. A value of 1.54 was used for the specific gravity of the pulp fibre.

The normal growth of micro-organisms in organic fraction consisting of about 53% carbon, 29% oxygen, 12% nitrogen and 6% hydrogen by weight (McKinney 1962). An approximate empirical formulation gives $C_5H_7O_2N$, which served as a guide in the selection of nutrient quantities. Pulp fibre, used as the organic material, supplied the carbon. Ammonium chloride (NH_4Cl) was used to supply nitrogen. Other nutrients found in bacterial protoplasm and their source compounds included K_2HPO_4 for phosphorus and potassium, $MgSO_4$ for magnesium, $CaCl_2$ for calcium and $FeCl_3$ for iron. About 1% (dry weight basis) of a municipal sludge provided seed sources for the anaerobic microbial species.

2.2 Sample Preparation

Pulp fibre board, after freeze drying, was separated into a fluffy mass. Selected proportions of dry kaolinite and dry fibre were mixed, and distilled water was added in amounts needed to form a slurry. Nutrients and seed material, in predetermined proportions, were added directly to the slurry for those experiments involving decomposition. Nitrogen served as the base by which all other nutrient quantities were selected. Approximate ratios giving optimal rates of decomposition included C/N = 30:1, P/N $\approx$ 1:5 and (Mg, Fe, K, Ca and Na)/N $\approx$ 1:16. These ratios appeared to satisfy the bacteria nutritional requirements and permitted almost complete degradation of the pulp fibres.

2.3 Consolidation Test

One-dimensional compression tests with pressures ranging from 122 kN/m^2 to 35 MN/m^2, were conducted on fibre-clay samples containing no nutrients using a special compression test cylinder (Fig. 1) with sample cross-sectional areas of 645 and 100 mm^2. The smaller area cylinder permitted higher pressures using the available equipment. Cylinder heights were sufficient to permit the use of fibre-clay samples with an initial slurry consistency. Water drainage through porous stones at the top and bottom provided a check on decrease in sample height. Only initial and final dial readings were taken for each load increment. A force transducer mounted below the bottom porous stone permitted load measurements to an accuracy of ±5 N.

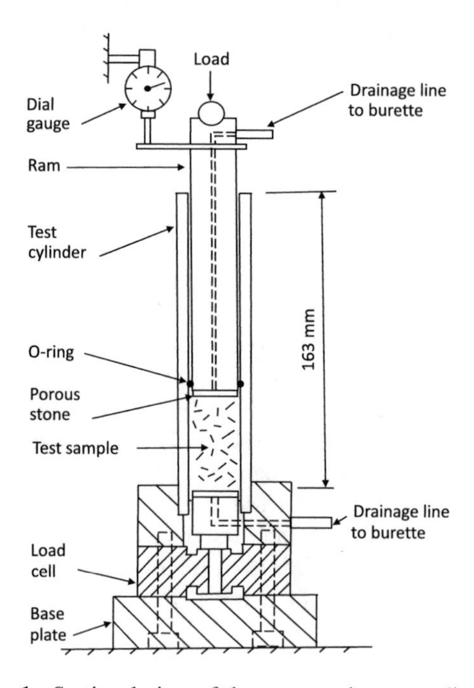

Fig. 1. Sectional view of the compression test cylinder

Slurry samples containing nutrients and seed material were consolidated at much lower pressures in a 3-L beaker with all drainage to the top porous stone (Fig. 2).

Consolidation loads of 0.47, 1.14, 2.28 and 3.42 kN/m^2 included three load increments-for the initial condition (no decomposition) and for later stages of partial decomposition. The second and third load increments were started as soon as the previous load increment reached 100% primary consolidation. A small vacuum at the top of the porous stone' removed fluids during drainage so as to avoid buoyancy. Drained fluids were returned to remoulded samples after each test series so that anaerobic decomposition could continue during storage at a temperature close to 35 °C.

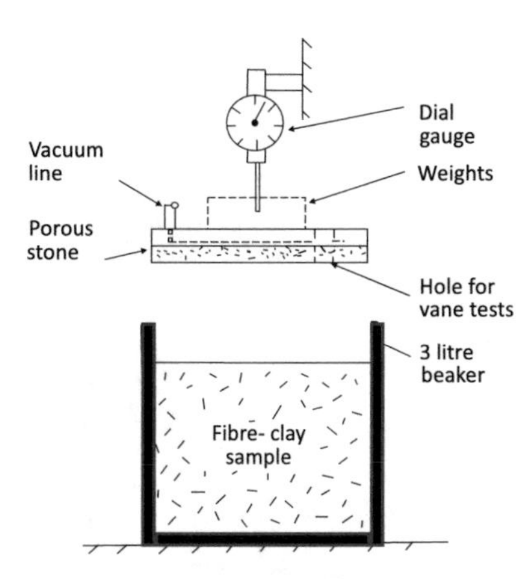

Fig. 2. Consolidation apparatus with a porous stone, weights and dial gauge

3 Soil Organic Content Determination

Decomposition involves microbial activity with the formation of gases, water and new bacterial cells, and a decrease in the organic solids content. Completion of the process means that the non-degradable residue (humus) material will remain. A brief review of the ignition test describes the method and the degree of decomposition.

3.1 Ignition Test

The direct determination of soil organic matter requires its separation from the inorganic solids. For engineering purposes, ignition of soil at high temperatures is the most common method. Destruction of the organic fraction by ignition requires that other soil constituents are not altered so that the weight loss, compared with the weight at 105 °C, can be taken as a measure of the organic content. Weight reduction curves for the soil components, kaolinite and pulp fibre, are summarized in Fig. 3.

The curves show that a temperature of 400 °C maintained for 12 h would result in a minimum loss of surface hydration water from the kaolinite. At higher temperatures, direct use of the ignition method without a correction factor would have given larger errors in organic content. Al-Khafaji and Andersland (1981), using experimental and published weight reduction curves for a number of clay minerals, concluded that the organic fraction X_f for the fibre-clay soils is given by:

$$X_f = 1 - X_m = 1 - C\frac{W_2}{W_1} \tag{1}$$

Where X_m is the mineral fraction, W_1 is the initial oven dry weight at 105 °C and W_2 is the final sample weight after ignition. The correction factor C depends on the ignition temperature and mineral soil type. Al-Khafaji and Andersland (1981) determined that C = 1.014 at 400 °C (12 h ignition) or C = 1.168 at 900 °C (1.5 h ignition) gave accurate values for the fibre-clay soils.

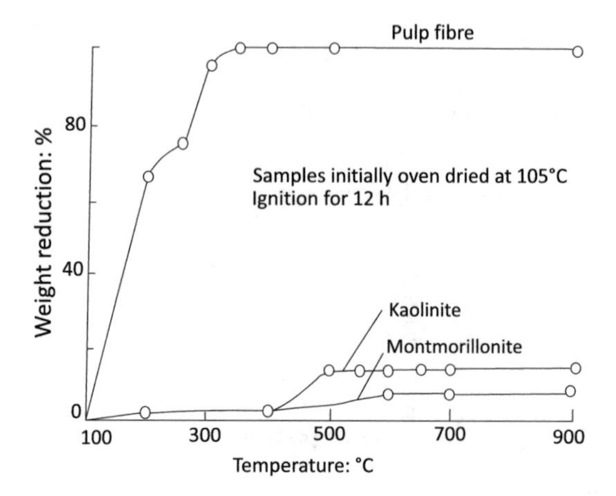

Fig. 3. Weight reduction against temperature curves (ignition for 12 h) for pulp fibre, kaolinite and montmorillonite (after Al-Khafaji 1979)

For comparison Arman (1970) recommended that soil organic content should be determined by combustion at 440 °C. Franklin et al. (1973) found that burning at 400–450 °C until constant weight is reached gave reliable organic contents. Skempton and Petley (1970) concluded that oven drying of organic soils at 105 °C and ignition of the organic material at 500 °C gave reasonable organic contents. They used a correction factor of 1.04 at 550 °C compared with C = 1.014 in Eq. (1) at 400 °C. Two reasons for the difference include the higher ignition temperature (550 °C against 400 °C) and their assumption that the organic carbon comprised 58% of the organic content. Broadbent (1965) stated that the carbon fraction (0.58) is based on very old work and the value may range from 0.5 to 0.58.

3.2 Decomposition Measurement

Decomposition in organic soils involves a breakdown and conversion of plant and animal residues into stable humus-like products. In predicting future soil properties, questions arise as to how much more decomposition may occur. The changes for a typical organic soil before and after partial decomposition are shown in Fig. 4.

It is reasonable to take the weight losses shown as a quantitative measure of the degree of decomposition X_{di}, thus

$$X_{di} = \frac{W_{fo} - W_{fi}}{W_{fo}} \tag{2}$$

where

W_{f0} = the initial weight of organic matter

W_{fi} = the weight of undecomposed matter after partial decomposition

Separation of the organic material from the inorganic fraction is not as simple as it first appears. Define the initial organic fraction $X_{fo} = W_{fo}/\left(W_{fo} + W_m\right)$ and the organic

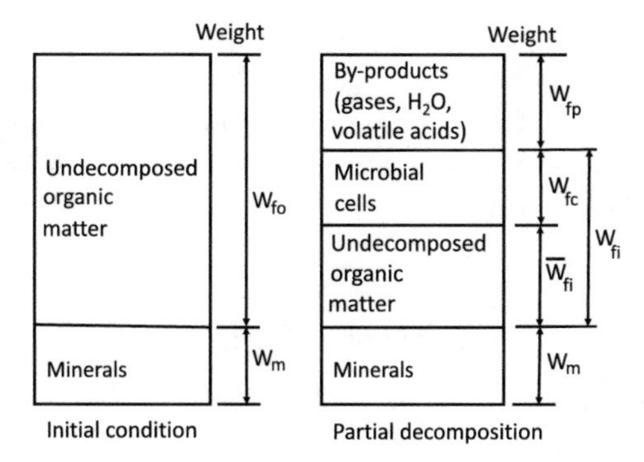

Fig. 4. Phase diagram showing organic soil, before and after partial decomposition (after Al-Khafaji, 1979)

fraction after partial decomposition $X_{fi} = \bar{W}_{fi}/\left(\bar{W}_{fi} + W_m\right)$ where symbols are as defined in Fig. 4. Solve for W_{fo} and $\bar{W}_{fi}$, substitute into Eq. (2) and with rearrangement write

$$X_{di} = \left(1 - \frac{X_{fi}}{X_{fo}}\right)\frac{1}{\left(1 - X_{fi}\right)} \tag{3}$$

Determination of X_{fo} and X_{fi}, using the ignition test, permits computation of the degree of decomposition X_{di}. In the usual ignition test, for determination of organic content, microorganisms are lumped with undecomposed organic material, hence a small correction for cell yield can be included in Eq. (3). Because this correction is small it has been omitted here. For more information on cell yield and a correction factor the reader is referred to Al-Khafaji (1979).

4 Soil Compressibility

Compression of soils at stress levels encountered in most engineering problems arises almost entirely as a result of a reduction in void volume. For decomposing organic soils, an additional factor is the loss of organic solids. Soil compressibility is discussed in terms of the fibre- clay soil decomposition effects.

4.1 Fibre-Clay Soil Behavior

The application of an external stress to a fine-grained soil causes a gradual decrease in water content until an equilibrium value is reached. The presence of organic material, with greater water holding capacity, greatly increases the equilibrium water content and the corresponding void ratio at low stress levels. This fact is illustrated in Fig. 5, where the organic fraction has been plotted against void ratio at selected pressures. For a pressure of 5 kN/m^2 the equilibrium void ratio in Fig. 5 ranges from 1.2 for a zero organic fraction to about 14 for an organic fraction close to unity. The influence of organic content is decreased with higher pressure levels and becomes very small for pressures greater than 8 MN/m^2.

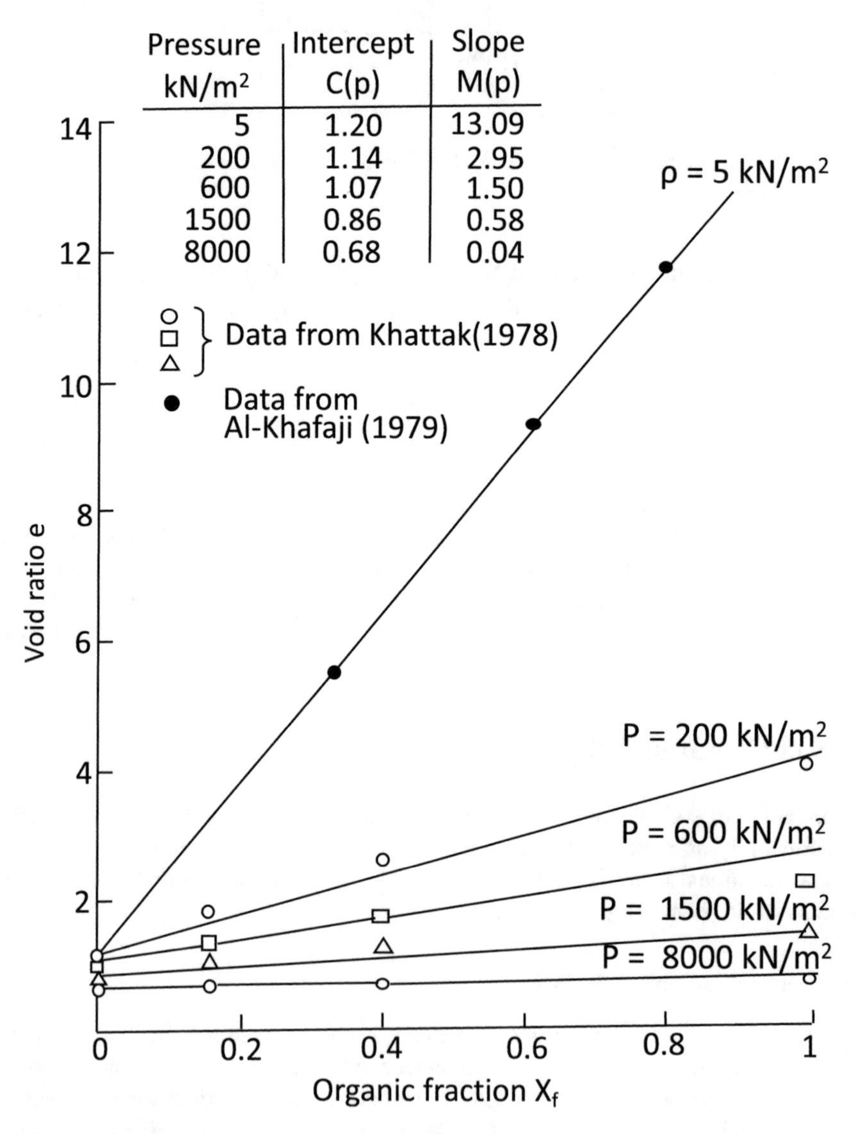

Fig. 5. Relationships between the organic fraction X_f, void ratio e and consolidation pressure p for the model soil samples

For an organic soil the equilibrium void ratio is dependent on composition, pressure, temperature and environment. The experimental data summarized in Fig. 5 represent a relatively constant temperature and chemical environment.

The data represent void ratios for one-dimensional compression and 24 h loading periods. For 200 kN/m^2 and higher pressures the data were obtained using the compression test cylinder shown in Fig. 1. Data for the 5 kN/m^2 pressure were obtained independently using 3 L samples and the apparatus shown in Fig. 2. For each pressure level an equation may be written giving the void ratio in terms of the organic fraction X_f. The intercept and slope of each line appear to be unique to the pressure. The linear relationships clearly satisfy the experimental data over the pressure range 5 kN/m^2 to 8 MN/m^2. The straight-line relationships in Fig. 5 show that the effects of organic content and consolidation pressure on void ratio can be uncoupled. The slope parameter $M(p)$ appears to be critical to equilibrium void ratio prediction as it experiences large changes with change in pressure. The intercept $C(p)$ undergoes less change with increase in pressure. Using the slope and intercept parameters, an equilibrium void ratio $e(p, X_f)$ may be computed as

$$e(p, X_f) = C(p) + M(p)X_f \tag{4}$$

The slope M(p) and intercept C(p) parameters are approximated using regression analysis as a function of the applied pressure, p in kN/m^2 follows

$$M(p) = \frac{24.85}{p^{0.3406}} - 1.266 \tag{5}$$

The correlation coefficient for this fit is $R^2 = 0.9991$. Similarly, the slope is related to the applied pressure p in kN/m^2 as

$$C(p) = \frac{1870 + 0.5679p}{1541 + p} \tag{6}$$

The correlation coefficient for this fit is $R^2 = 0.9848$. Thus, the slope and intercept given by Eqs. (5) and (6) are shown as functions of pressure in Fig. 6

Note that for a given pressure p_o the corresponding equilibrium void ratio $e_o(p_o, X_f)$ has been determined for all organic fractions. Comparative results for Eq. (4) have been difficult to obtain because void ratios reported in the literature are generally given without information on effective overburden pressures and organic fractions. The paper mill sludge resembles the fibre–clay soils in that it consists primarily of kaolinite and waste pulp fibres. Using soil properties reported for the block samples, three void ratios were computed. Ash contents were converted to organic fractions using Eq. (1) with a correction factor for loss of surface water associated with the 900 °C ignition temperature (Charlie 1975).

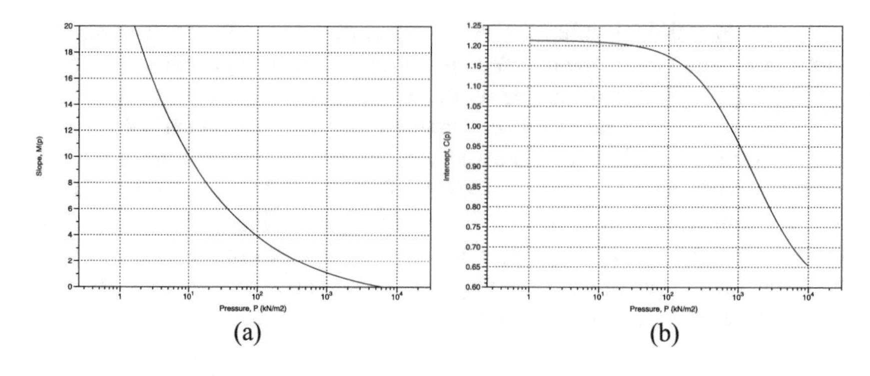

2

Fig. 6. The Slope (a) and Intercept (b) Parameters versus pressure, P (kN/m^2)

4.2 Fibre-Montmorillonite Clay and Fibre-Sand Soil Behaviour

Compression of soils at stress levels encountered in most engineering problems arises almost entirely as a result of a reduction in void volume. The procedures just outlined was employed to study the behaviour Fibre-montmorillonite clay soil (Tian 1991). This study didn't include decomposition effects. The results indicate that a linear relationship exists between the void ratio and the organic fraction under constant pressure. Similar relationships were observed for Sand-fibre mixtures. However, these studies are beyond the scope of this paper.

4.3 Settlement Prediction Due to Surface Loads

One-dimensional volume change, caused by surface loads on fibre-clay soil deposits, can be predicted using equilibrium void ratios given by Eq. (4) and soil mechanics theory. Both the initial void ratio e_o at a pressure p_o and the final void ratio e_f at a pressure $p_f = (p_o + \Delta p)$ will be dependent on the initial organic fraction X_{fo}. The settlement S_i for a soil layer of initial thickness Δz_i can be expressed in terms of Eq. (4) and the change in void ratios as

$$S_i = \Delta z_i \left(\frac{e_0 - e_f}{1 + e_0} \right) = \Delta z_1 \left[\frac{C(p_0) - C(p_f) + [M(p_0) - M(p_f)] X_{f0}}{1 + C(p_0) + M(p_0) X_{f0}} \right] \quad (7)$$

Equation (7) is valid regardless of the mechanism causing volume change-primary consolidation, secondary compression or loss of organic solids by decomposition. The total settlement S_T can be obtained by summing incremental settlements over the thickness of the soil deposit, thus $S_T = \sum S_i$. For natural soil deposits, the organic fraction will typically vary with depth owing to differences in deposition history and degree of decomposition. These variations in organic content can be accounted for by proper selection of the depth increment.

Example 1 A 5 m thick soil layer with an organic fraction X_{fo} equal to 0.25 has an average initial effective overburden pressure p_o equal to 5 kN/m^2. Determine the settlement for a load increment of 95 kN/m^2.

Solution 1. Using Eqs. (5) and (6) computer the parameters $C(p_o)$ and $M(p_o)$ for the loads p_o and $P_f = (5 + 95) = 100$ kN/m^2 as follows

$$p_o = 5\,\mathrm{kN/m^2} \qquad C(p_o) = 1.20 \qquad M(p_o) = 13.097$$

$$p_f = 100\,\mathrm{kN/m^2} \qquad C(p_f) = 1.17 \qquad M(p_f) = 3.912$$

Substitution into Eq. (7) gives the settlement

$$S_i = (5\,\mathrm{m})\frac{1.20 - 1.17 + (13.097 - 3.912)(0.25)}{1 + 1.20 + 13.097(0.25)} = 2.125\,\mathrm{m}$$

The predicted settlement includes primary consolidation and secondary compression associated with the 24 h laboratory loading period on a small fibre-clay soil sample

4.4 Settlement Prediction Due to Decomposition

Settlement due to loss of organic solids, brought about by decomposition, may in some cases be more significant than settlement due to higher surface loads. A method for dealing with decomposition of organic solids (Al-Khafaji 1979) requires use of a height ratio factor ψ defined as

$$\psi = \frac{\textit{organic solids volume}}{\textit{total solids volume}} = \frac{X_{f0}}{\frac{G_{sf}}{G_s} + \left(1 - \frac{G_{sf}}{G_s}\right)X_{f0}} \tag{8}$$

where G_{sf} is the specific gravity of the organic solids, G_s that of the mineral solids and X_{fo} is the initial organic fraction. The decomposed layer thickness Δz_d, computed in terms of ψ, becomes

$$\Delta z_d = \frac{1 + e_f}{1 + e_0}(1 - \psi X_{di})\Delta z_i \tag{9}$$

where Δ_{zi} is the initial layer thickness, e_o and e_f are the initial and final void ratios, and X_{di} is the degree of decomposition. For the fibre-clay soils the mineral fraction (kaolinite) had a specific gravity of 2.70 and the organic fraction (fibre) a value of 1.54. The following example illustrates the use of Al-Khafaji's (1979) method.

Example 2 A 5 m thick soil layer with an initial organic fraction X_{fo} equal to 0.35 has an average initial effective overburden pressure p_o equal to 5 kN/m^2. It is anticipated that about 25% decomposition (X_{di} = 0.25) will occur during the time interval under consideration. Assume an initial void ratio of 5.13, then calculate the final the soil layer thickness.

Solution 2. Compute the decomposed organic fraction X_{fi} from Eq. (3) as follows

$$X_{fi} = \frac{X_{f0}(1 - X_{di})}{1 - X_{di}X_{f0}} = \frac{0.35(1 - 0.25)}{1 - (0.25)0.35} = 0.288$$

Also, using Eq. (4) yields the final void ratio

$$e_f = C(p_0) + M(p_0)X_{fi} = 1.20 + 13.097(0.288) = 4.970$$

Using Eqs. (8) and (9) with G_{sf} = 1.54 and G_s = 2.70 compute

$$\psi = \frac{X_{f0}}{\frac{G_{sf}}{G_s} + \left(1 - \frac{G_{sf}}{G_s}\right)X_{f0}} = \frac{0.35}{\frac{1.54}{2.70} + \left(1 - \frac{1.54}{2.70}\right)(0.35)} = 0.486$$

$$\Delta z_d = \frac{1 + e_f}{1 + e_0}(1 - \psi X_{di})\Delta z_i = \frac{1 + 4.86}{1 + 5.14}[1 - (0.486)(0.25)](5) = 4.192\,\text{m}$$

Settlement due to decomposition for a constant effective overburden pressure now equals the difference, thus

$$S_d = \Delta z_i - \Delta z_d = 5.00 - 4.192 = 0.807\,\text{m}$$

Compression due to an increase in load requires new void ratio values based on the decomposed organic fraction X_{fi} and the load increment.

5 Conclusions

The use of compression index in the computation of settlement of foundations in organic soil deposits is invalid. This paper offers a new technique for computing the settlement of organic soils without the need for compression index. The effects of decomposition on the compressibility of fibre-clay soils prepared from kaolinite and pulp fibres was studied. These conclusions are intended to reflect the findings of this investigation and are limited to the materials used and the test procedures employed. Additional testing for other types of soil-fibre mixture are needed to develop a comprehensive overview of the new method presented in this paper.

The author has established that the relationship between void ratio and organic fraction (by weight) for a constant stress level is linear. A plot of the intercepts C and slopes M against pressure p gave the relationship between p and the new void ratio parameters $C(p)$ and $M(p)$. This allows the decoupling the effects of organic content and pressure on the equilibrium void ratio. Use of these parameters with the organic fraction X_f permits prediction of an equilibrium void ratio $e(p, X_f)= C(p)+ M(p)X_f$ for the fibre-clay soils mixtures. Similar behaviour was observed for sand and montmorillonite soil mixtures with organic content greater than 30% by weight.

Ultimate settlement in fibre-clay soils can be predicted in terms of the organic fraction and increase in pressure using the equilibrium void ratios and conventional soil mechanics theory. Increased settlement, as a result of decomposition with loss of organic solids volume, can be predicted using the initial and decomposed organic fractions with effective overburden pressures. A simplified computation procedure is introduced to assess the magnitude of settlement for a given degree of consolidation.

Additional research on the behaviour of fiber-clay mixtures with different types of minerals and organic matter is required to help develop more accurate relationships between the void ratio, applied pressure and organic content for all soils.

Acknowledgments. I am grateful to my late Professor Orlando B. Andersland for showing me the joy of scholarship and research. His guidance and support will not be forgotten.

References

AI-Khafaji, A.W.N.: Decomposition effects on engineering properties of fibrous organic soils. PhD thesis, Michigan State University (1979)

AI-Khafaji, A.W.N., Andersland, O.B.: Ignition test for soil organic-content measurement. J. Geotech. Engng Div. Am. Soc. Civ. Engrs **107**(GT 4), 465–479 (1981)

Arman, A.: Engineering classification of organic soils. Highw. Res. Rec. **310**, 75–89 (1970)

Broadbent, F.E.: Organic matter. In: Methods of Soil Analysis, Part 2, Chemical and Microbiological Properties, Chapter 92; No. 9 in the Series Agronomy. Wisconsin: American Society of Agronomy, Madison (1965)

Charlie, W.A.: Two cut slopes in fibrous organic soils, behavior and analysis. Ph.D thesis, Michigan State University (1975)

Franklin, A.G., Orozco, L., Semrau, R.: Compaction and strength of slightly organic soils. J. Soil Mech. Fdns Div. Am. Soc. Civ. Eng. **99**(SM7), 541–557 (1973)

Golueke, C.G.: Composting. Rodale, Emmaus, Pennsylvania (1972)

McKinney, R.E.: Microbiology for Sanitary Engineers. McGraw-Hill, New York (1962)

Mesri, G.: Engineering properties of fibrous peats. J. Geotech. Geoenviron. Eng. **133**(7) (2007)

Perkin-Elmer Corporation: Instruction manual of Perkin-Elmer shell model 212B sorptometer. Norwalk, Connecticut: Instrument Division, Perkin-Elmer Corporation (1961)

Skempton, A.W., Petley, D.J.: Ignition loss and other properties of peats and clays from Avonmouth. King's Lynn and Cranberry Moss. Géotechnique **20**(4), 343–356 (1970)
Tian, L.: Volume Change Behavior of Model Organic Soils. MS thesis, Bradley University (1991)

Resilient Modulus Prediction of Subgrade Soil Using Dynamic Cone Penetrometer

Aneke Frank Ikechukwu[1(✉)], Ojiogu Emeka[2], and Mostafa M. Hassan[1]

[1] Sustainable Urban Road Transportation (SURT) Research Group, Department of Civil Engineering and Information Technology, Central University of Technology, Private Bag X20539, Free State, Bloemfontein 9300, Republic of South Africa
frankaneke4@gmail.com, mmostafa@cut.ac.za
[2] Department of Civil Engineering, Enugu State University of Science and Technology (ESUT), Independence Layout/P.M.B, Enugu 01660, Nigeria

Abstract. Stress-strain response of pavement structure is commonly presented with resilient modulus (M_r). Resilient behavior of subgrade soils is measured using expensive laboratory tests, that is somewhat time-consuming. Design engineers sometimes, uses overestimated backcalculated values from non-destructive test, which in turn lead to over-design of pavement. These challenges encourage the need for a valuable and inexpensive insitu geotechnical testing procedures that can easily and directly determine M_r of subgrade. As most of this testing equipment are not readily available in the University laboratories and highway engineering research centers in Nigeria. Furthermore, it is very difficult to perform M_r testing at desired in-situ condition. Therefore, it becomes imperative to evaluate M_r of subgrade soils using dynamic cone penetrometer (DCP) and California Bearing Ratio (CBR) Test. Such a rational approach is used in this investigation to ascertain resilient performance of an existing pavement foundation (subgrade) situated at Enugu - Abakaliki – Ikom Highway Road, in South-East of Nigeria. The experimental program includes tests, routinely used in geotechnical engineering such as: Particle size distribution, Atterberg Limits, compaction, CBR, DCP and M_r. The interpretation of test results confirmed, DCP as a useful tool to determine CBR, resilient and unconfined compressive strength (UCS) values of an in-situ pavement structure. The predicted CBR values from AFCP-LVR software after DCP testing shows that laboratory soaked CBR value is averagely 1.6% lower. The predicted M_r values using AFCP-LVR software indicted a variation of 5.7 MPa on the average compared to measured M_r values.

1 Introduction

Pavement performance depends on the mechanical properties of the subgrade. M_r is the major design parameter commonly used for structural analysis of pavement (AASHTO 2003, M-EPDG). M_r is measured using cyclic triaxial cell on representative specimens. Though, DCP can be used to estimate soil strength for studies related to design and strengthening of existing pavement Scala (1956). Mechanistic-empirical (M-E) method, considers M_r a stiffness parameter required to analyze fatigue on pavement

S. Hemeda and M. Bouassida (Eds.): GeoMEast 2018, SUCI, pp. 67–87, 2019.
https://doi.org/10.1007/978-3-030-01941-9_6

structure. Despite, deviator and confining stress, are stress state variables that controls M_r under cyclic condition (Li and Selig 1994). Though, there are other variables that influence the stress variables of subgrade and these are: soil gradation, compaction method, specimen size and testing procedure.

Hassan (1996) developed first ever M_r-DCPI correlation and discovered that the relationship was more significant at the optimum moisture content (OPT) than wet of optimum (WOP). Despite the coefficient of determination is >0.80

$$M_r = 7013.065 - 2040.783 * \ln(\mathrm{DCPI}),\ R^2 = 0.41 \tag{1}$$

Where:

M_r (MPa) = resilient modulus
DCPI (mm/blow) = field dynamic cone penetrometer index.
R^2 = coefficient of determination

Jianzhou et al. (1999) proposed a correlation obtained from back calculation of the falling weight deflectometer, the values of M_r predicted with this method, was too low compared to the laboratory measured procedures, with lower coefficient of determination ($R^2 = 0.42$). Thus, the scholars maintained that the validity of their equation is not based on R^2, rather on the applicability and accuracy.

$$M_r = 338 * (\mathrm{DCPI})^{-0.39} \tag{2}$$

Where:

M_r (MPa) = resilient modulus
DCPI (mm/blow) = field dynamic cone penetrometer index.
R^2 = coefficient of determination

George and Uddin (2000) reported on the manual and automated DCP. Hence, they discovered that there was no significant difference between these two types of DCP, despite having low coefficient of correlation value.

$$M_r = 532 * (\mathrm{DCPI})^{-0.492},\ R^2 = 0.40 \tag{3}$$

Where:

M_r (MPa) = resilient modulus
DCPI (mm/blow) = field dynamic cone penetrometer index.
R^2 = coefficient of determination

Gudishala (2004) correlated field DCPI, dry density and moisture content (M_c) and stated that M_c greatly influence M_r of subgrade soil.

$$M_r = \frac{1100 * (\mathrm{DCPI})^{-0.44}}{M_c} + 2.39 * (\gamma_d),\ R^2 = 0.68 \tag{4}$$

Herath et al. (2005) proposed two equations using DCPI and the second equation was correlated using DCPI, dry density (γ_d) and plasticity index (PI). These equations were strongly proved reliable, but they declared Eq. 6, suitable for all soil type and more reliable than Eq. 5. Despite Eq. 6 coefficient of determination been less than 0.80.

$$M_r = 16.28 + \frac{928.24}{DCPI}, \; R^2 = 0.82 \tag{5}$$

$$M_r = 520.62 * (DCPI)^{-0.738} + 0.4 * \left(\frac{\gamma_d}{M_c}\right) + 0.44 * (PI), \; R^2 = 0.78 \tag{6}$$

Mohammed et al. (2007) propounded two equations that are in line with Eqs. 5 and 6 with strong correlation values of 0.91 and 0.92, which after their studies also claimed that their equation is for all type of fine-grained subgrade soil and granular materials.

$$M_r = 151.8 * (DCPI)^{-1.096}, \; R^2 = 0.91 \tag{7}$$

$$M_r = 165.5 * (DCPI)^{-1.147} + 0.097 * \left(\frac{\gamma_d}{M_c}\right), \; R^2 = 0.91 \tag{8}$$

Where:

γ_d = dry density
$M_c(\%)$ = Moisture content
M_r (MPa) = resilient modulus
R^2 = coefficient of determination
DCPI (mm/blow) = field dynamic cone penetrometer index.

Nonetheless, DCP applications are discovered to be limited to the field conditions as a result of confining effects caused by heavy mass of hammer. With the intentions of overcoming this issue and to expand the application of DCP in the laboratory. Nguyen and Mohajerani (2012) proposed lightweight dynamic cone penetrometer with a hammer of 2.25 kg and a falling height of 510 mm. The results from their experimental works, found the following relationship between M_r and the lightweight dynamic penetrometer index (DLPI):

$$Log(M_r) = 2.242 - 0.890 Log(DLPI), \; R^2 = 0.64 \tag{9}$$

Where:

M_r (MPa) = resilient modulus
DCPI (mm/blow) = dynamic cone penetrometer index.

Though, other empirical correlations exist for determination of M_r that are based on soil static properties such as CBR. However, CBR does not represent cyclic response of pavement materials under moving vehicles. In addition, over prediction "of resilient modulus, could be responsible for over-design and as such will render the design to be costly.

This paper presents laboratory and field DCP testing programs performed to investigate the applicability of DCP as a tool to evaluate resilient behaviour of an existing 1.8 km long highway pavement. Thus, establishing a mathematical correlation between M_r and DCP, and CBR and DCP is beyond the scope of this study.

2 Methodology

2.1 Field Work

The DCP tests was conducted according to ASTM-D6951-3 (2003) testing procedures, the apparatus consists of 16 mm diameter steel rod in which a tempered steel cone of 20 mm diameter base and 600^{-point} angle is fixed. The DCP was driven into the soil by an 8 kg hammer at free fall of 575 mm and DCP index was calculated. The Field evaluation was carried out on subgrade soil of the pavement Fig. 1. Penetration depth of 800 mm was adopted (Fig. 2), because the induced stresses due to wheel load becomes reduced beyond this depth. The road was subdivided into 6 test sections for this study and at each section average of three cores were drilled. Test pits were excavated at the edge of pavement were crack and fatigues were identified, as to obtain detailed pavement profile and causes of those distress. Field moisture content was determined using stir-frying according to ASTM D 4959 (2000). While the dry density of subgrade was measured using drive cylinder test in accordance with ASTM D2937 (2004).

Fig. 1. Drilled cores near fatigue area

2.2 Laboratory Work

The soils used in this study were sampled from Enugu - Abakaliki – Ikom Highway Road, located in South-East of Nigeria on x, y coordinate of latitude 06.11562°N and 007.79615°E with GPS precision of 0.4 m away from sampling site. The soils were obtained close to the drill cores at depths ranging from 0.6 to 1.2 m below ground surface, very close to the damaged structure where fatigues and cracks were observed. Summary of the soils classifications are presented in Table 1. The soil from the subgrade was collected, bagged and sealed to avoid moisture lost within the soil pore and transferred to Enugu State University of Science and Technology (ESUT) Civil Engineering Laboratory for detailed geotechnical testing such as gradation, consistency, Proctor compaction, natural moisture content, CBR and resilient modulus test in accordance with American Standard for Testing and Materials (ASTM). The CBR testing was carried in other to validate the estimated values from AFCP-LVR software.

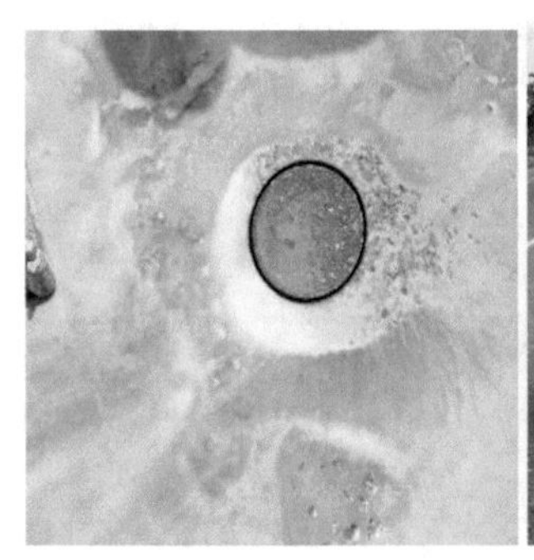

Fig. 2. Dynamic cone penetrometer test

Table 1. Grain size analysis and index properties of the investigated soils

Index property	Seg. 1	Seg. 2	Seg. 3	Seg. 4	Seg. 5	Seg. 6
Sand/gravel	72.95	69.22	56.44	51.44	74.46	65.94
Silt	13.03	12.19	18.13	21.15	10.14	14.91
Clay	14.021	18.59	25.43	27.41	15.40	19.15
D_{10}	0.012	0.015	0.008	0.021	0.041	0.035
D_{30}	0.11	0.07	0.03	0.04	0.09	0.05
D_{60}	0.21	0.25	0.15	0.11	0.15	0.25
C_u	16.67	18.67	18.75	5.08	3.75	7.14
C_c	4.17	1.31	0.75	0.81	1.35	0.29
W_n (%)	17.12	18.28	18.94	17.91	18.05	18.17
Specific gravity, G_s	2.67	2.73	2.68	2.70	2.69	2.72
Liquid limit (%)	28.14	29.78	28.10	30.09	30.16	28.47
Plastic limit (%)	16.78	18.34	16.56	15.76	15.52	16.48
Plastic index (%)	11.36	11.44	11.54	14.33	14.64	11.99
AASHTO	A-2-6	A-2-6	A-2-6	A-2-6	A-2-6	A-2-6
Shrinkage limit (%)	9.61	8.26	9.81	9.15	9.06	8.54
OPT (%)	14.87	15.41	15.45	15.38	15.60	14. 65
MDD (kN/m^3)	21.80	22.60	22.48	22.23	22.30	21.96

[a]Natural moisture content (W_n)
[b]Optimum moisture content (OPT)
[c]Maximum dry density (MDD)

3 AFCP-LVR Software

AFCP-LVR software is a South African software developed at the Council for Scientific and Industrial Research (CSIR). Through a collaborative project, this software was commissioned by AFCAP between the CSIR South Africa and experienced Low Volume Road (LVR) practitioners. The original Win DCP software developed by CSIR has now been upgraded into a user-friendly software tool specifically tailored to upgrading of LVRs and high-volume road (HVL) to a paved standard using the Dynamic Cone Penetrometer (DCP). It incorporates automated procedures for analyzing DCP field data from the existing pavements as well as for laboratory testing and evaluation of imported materials as inputs for the design of new pavement layers (Table 2).

Table 2. Measured CBR Results

CBR	Seg. 1	Seg. 2	Seg. 3	Seg. 4	Seg. 5	Seg. 6
Average moisture content for 4 days Soaked (%)	17.30	17.11	17.97	17.56	18.04	18.12
Average moisture content for 4 days unsoaked (%)	14.60	14.12	14.53	14.31	13.91	14.28
CBR values 4 days soaked (%)	37.41	35.19	49.13	36.18	46.35	47.56
CBR values 4 days unsoaked (%)	62.15	67.21	63.19	60.34	65.21	66.31

4 Results Interpretation

4.1 Sieve Analysis

The soils tested for particle size distribution was segmented into 6 parts and soil samples were collected at the interval of 300 m, by drilling 3 different cores at each section. The set of sieves opening used for the sieve analysis ranges from 19 mm to 0.075 mm in descending order ASTM D6913 (2009). Figure 3 indicated primarily that the soils consisting mainly of gravel particles. The soils have minimum particles sizes ranging from 0.025 mm to 0.058 mm as determined using hydrometer analysis, whereas the coarse gravel for all the soils ranges from 4.92 mm to 18.76 mm. The soils fine content passing sieve size #200, varies at each segment i.e. Seg. 1 the fine content is 20.05%, Seg. 2 contain 30.77% fine, Seg. 3 possess 43.56%, while Seg. 4, 5 and 6 contained 48.56%, 25.54% and 34.05% fines respectively. The summary of the average particle size (D_{60}), effective size (D_{10}), uniformity coefficient (C_u), and coefficient of curvature (C_c) are summerized in Table 1.

4.2 Index Property

The principal aim of these test was to identify changes in plastic limit, liquid limit and plasticity index. The soil classification conducted according to AASHTO system, all

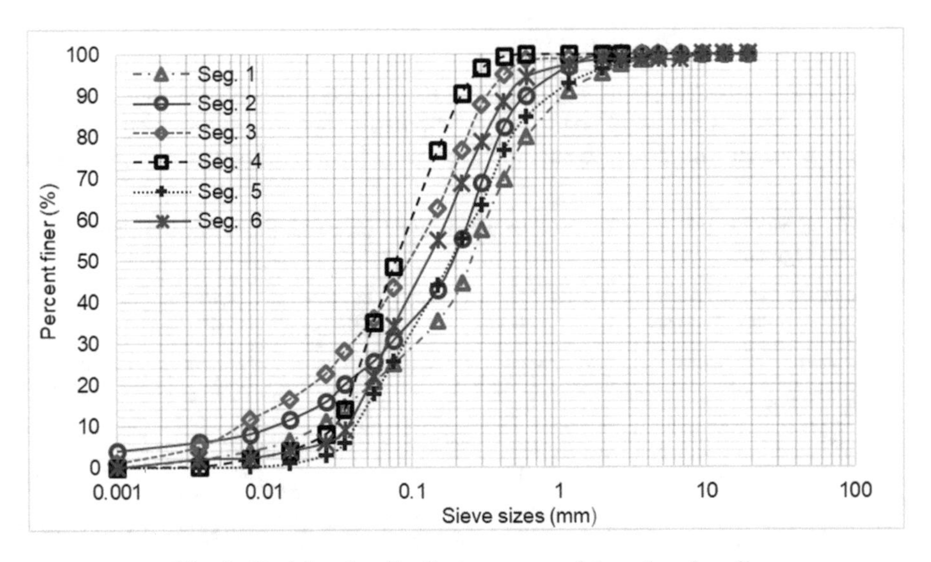

Fig. 3. Particles size distribution curves of the subgrade soils

the soils are classified as A-2-6. The detailed index properties of the soil samples are summarized in Table 1.

4.3 California Bearing Ratio

The CBR testing was carried out according to D4429. Measured CBR values were compared with the CBR values estimated from AFCP-LVR software Table 3. Findings from the summarized results of soaked CBR and unsoaked CBR for all soil segments indicated that unsoaked CBR values are greater than that of the soaked CBR in all tested soils. The CBR values of all the unsoaked samples ranges from 60.34% to 67.21% and this implies that the tested subgrade segments are excellent. Whereas, under soaked condition the CBR values of all tested samples ranges from 35.19% to 49.13% and this is considered good subgrade (Yoder and Witczak 1975).

4.4 Dynamic Cone Penetrometer (DCP)

Dynamic Cone Penetrometer is an in-situ test, that measures the strength of subgrade. During testing the penetration depth versus corresponding number of blow were collected, recorded and analyzed. Results of the overburden pressure, strength and stiffness of the base layer were determined. The software predicted the CBR values of the tested soil, which is 2.6% higher than the laboratory measured soaked CBR values on the average. However, corrections against operators, moisture content, and grain size were not applied during the course of conducting DCP testing. Figures 3, 4, 5, 6, 7, 8, 9 presents DCP curves of penetration depth versus number blows. The curves show that the penetration depth increases as the number of blows increase. This response of the subgrade with respect to number of blows can be attributed to the increased moisture content as the penetration depth deepens.

4.5 AFCP-LVR Software Analysis

AFCP-LVR software, extensively explains the condition of the existing pavement structure. Observations with road design, stipulated that a well-designed and constructed pavement over the years should be well balanced i.e. the strength should

Table 3. E-Moduli (MPa) and layer strength diagram (Existing Pavement Structure) for all pavement segments

Seg. 1 depth	Ave. estimated e-moduli at each depth	Measured M_r	E-moduli range (MPa)	Estimated CBR at each depth	Measured CBR	R^2	UCS
(mm)	(MPa)	(MPa)	5P - 95P	(%)	(%)		(kPa)
0 - 150	463	–	68 - -1	143	–		1185
151 - 300	902	–	236 - 4241	260	–		1999
301 - 450	686	–	191 - 2804	208	–		1646
451 - 600	147	–	37 - 762	187	–		355
601 - 800	97	89	47 - 323	49	37.41	0.835	292
Seg. 2							
0 - 150	491	–	84 - -1	153	–		1258
151 - 300	534	–	149 - 2156	170	–		1375
301 - 450	444	–	27 - 11713	120	–		346
451 - 600	128	–	28 - 604	154	–		156
601 - 800	105	97.32	28 - 193	48	35.19	0.845	170
Seg. 3							
0 - 150	417	–	69 - -1	126	–		1061
151 - 300	244	–	68 - 1005	67	–		603
301 - 450	180	–	60 - 548	46	–		439
451 - 600	102	–	27 - 459	23	–		241
601 - 800	101	98	33 - 414	39	36.18	0.963	264
Seg.4							
0 - 150	417	–	69 - -1	146	–		1061
151 - 300	244	–	68 - 1005	167	–		603
301 - 450	180	–	60 - 548	146	–		439
451 - 600	102	–	27 - 459	123	–		241
601 - 800	93	87	33 - 414	52	49.13	0.932	264
Seg. 5							
0 - 150	328	–	70 - 3342	95	–		823
151 - 300	111	–	28 - 595	26	–		263
301 - 450	69	–	24 - 202	15	–		159
451 - 600	60	–	21 - 177	13	–		138
601 - 800	113	108	21 - 197	51	46.35	0.832	145
Seg. 6							
0 - 150	202	–	58 - 783	153	–		493
151 - 300	101	–	31 - 341	123	–		238
301 - 450	77	–	27 - 226	117	–		180
451 - 600	60	–	19 - 193	112	–		137
601 - 800	92	87	18 - 152	50	47.56	0.884	118

deteriorate simultaneously with depth from the surface and this is referred to as strength balance. The change in strength of pavement layers with depth, is indicated as strength balance of the pavement structure. The smooth decrease in strength without discontinuities, is depicted as well balanced and DCP design analysis is based on this concept. Nevertheless, some of the pavement segments especially the ones very close to the identified cracks and fatigues section were discovered by the software to decrease in strength with discontinuities i.e. Seg. 1, 2, and 4 are poorly balanced deep structure (PBD). Whereas, Seg. 3, 5 and 6 are averagely balanced deep structure (ABD) because pavement around this segment decreases smoothly in strength without discontinuities. Pavement balance at any depth is determined from the formula proposed by Kleyn et al. (1989).

$$\mathrm{DSN}(\%) = \left\{\mathrm{D} * [400 * \mathrm{B} + (100 - \mathrm{B})2] / \left[4 * \mathrm{B} * \mathrm{D} + (100 - \mathrm{B})^2\right]\right\} \tag{10}$$

Where

DSN = pavement structure number (%)
B = parameter defining the standard pavement balance curve (SPBC)
D = pavement depth (%)

This equation accommodates series of developed curves for various pavement structure numbers and depths. This is presented in Standard Pavement Balance Curves (SPBC) as shown in Figs. 4, 5, 6, 7, 8, 9.

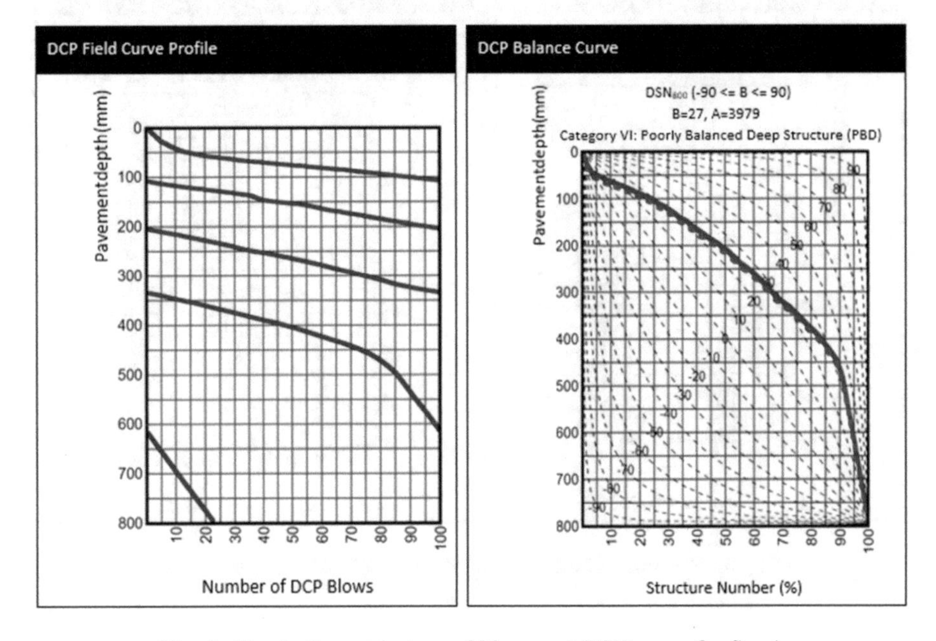

Fig. 4. Penetration against no of blows and DSN curve for Seg.1

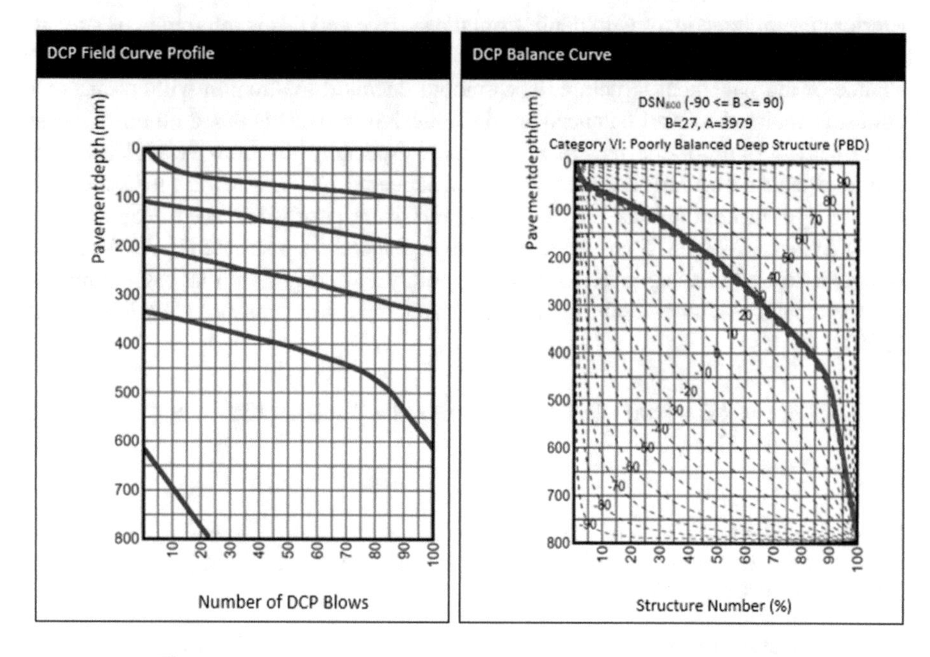

Fig. 5. Penetration against no of blows and DSN curve for Seg.2

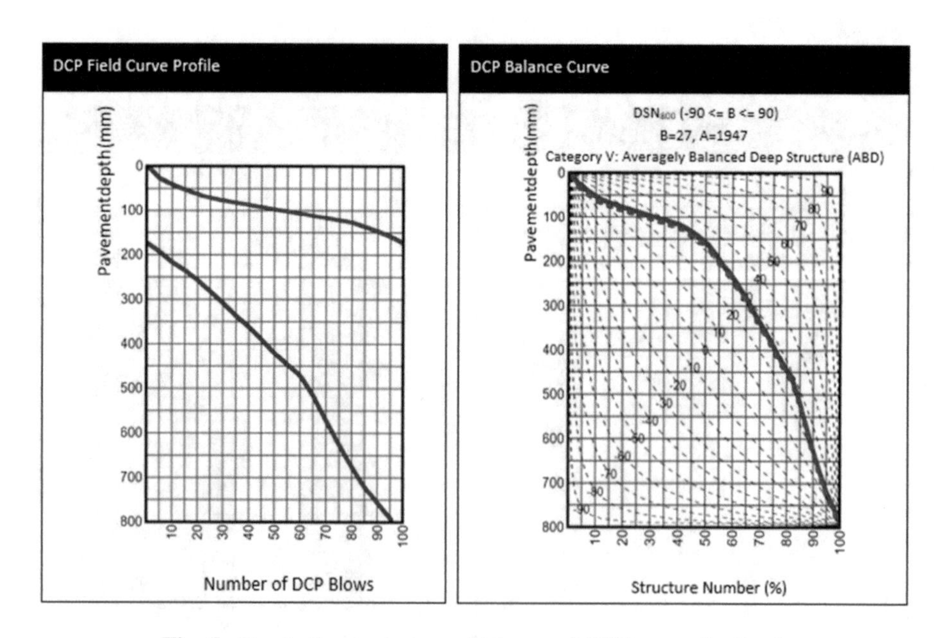

Fig. 6. Penetration against no of blows and DSN curve for Seg.3

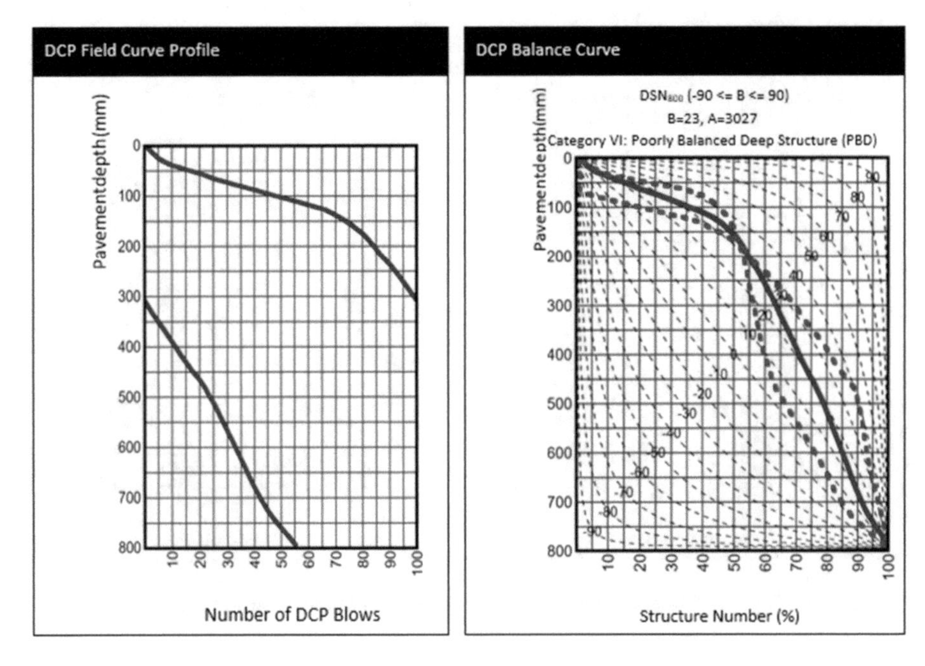

Fig. 7. Penetration against no of blows and DSN curve for Seg.4

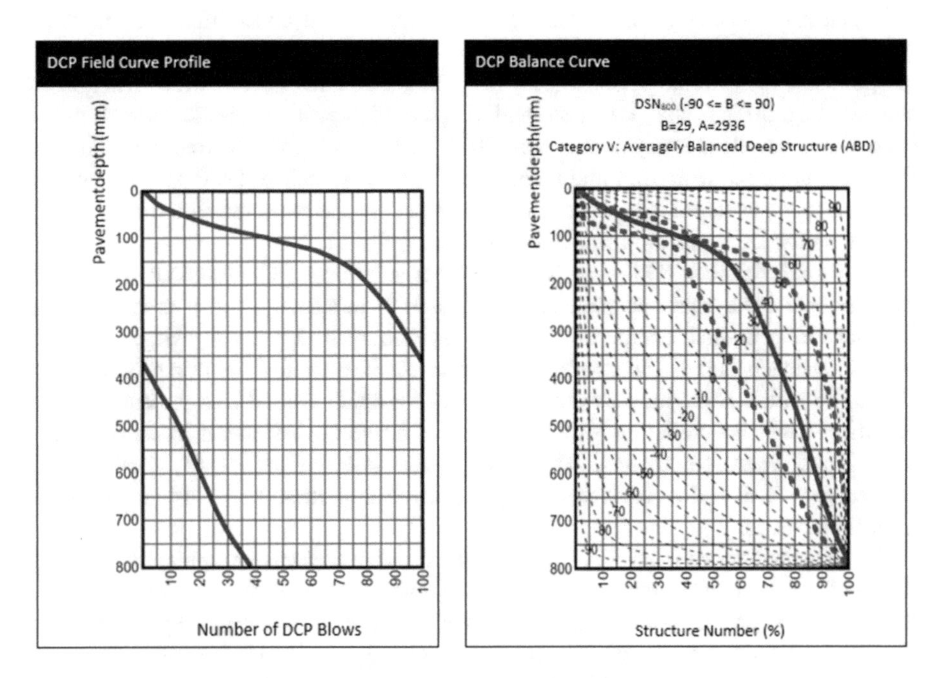

Fig. 8. Penetration against no of blows and DSN curve for Seg.5

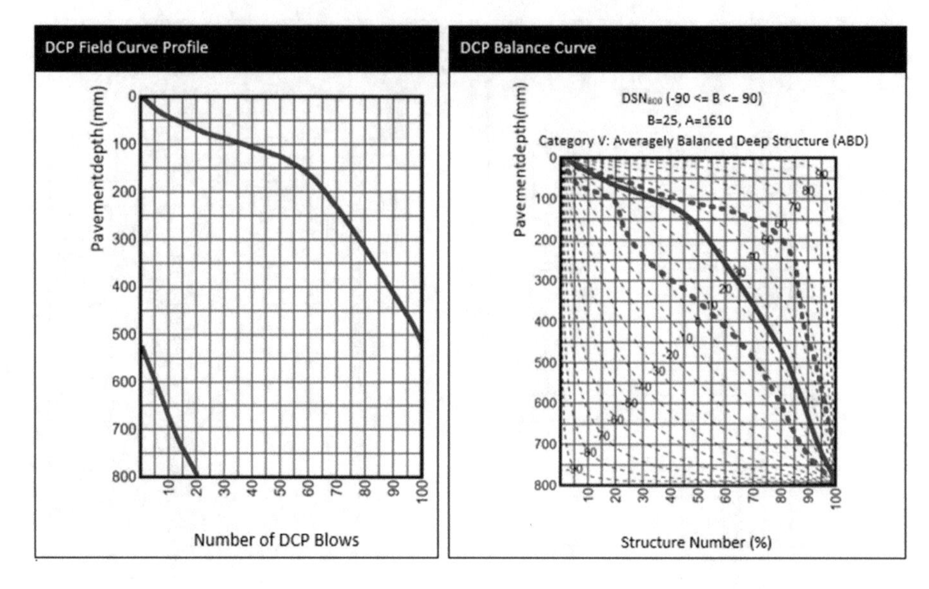

Fig. 9. Penetration against no of blows and DSN curve for Seg.6

4.6 Layer Strength Diagram

This phase in AFCP-LVR software gave explicit details concerning the structural behaviour of the existing pavement. Analysis of the DCP data gave adequate strength performance of various pavement layers, when fitted in the software using layer strength diagram interface. Layer strength diagram illustrates the in-situ strength of the pavement materials at every depth (Figs. 10, 11, 12, 13, 14, 15). The rate of penetration is then compared with minimum specified standards, called DCP master curves. Evaluation of pavement adequacy at various depth for the expected future traffic load, were used for DCP profiling Figs. 16, 17, 18, 19, 20, 21.

Structural Number (DSN_{800}) is the number of blows DCP required to reach a certain depth for a balanced pavement expressed in percentage and it is the Balance Number (BN) at that depth. classification system was developed using pavement strength balance curves. Moreover, the pavement is classified in terms of Balance Curve (B) which is the balance curve followed by the measured balance curve of the pavement and the deviation (A) between the Standard Pavement Balance Curve (SPBC) and the measured curve. The dotted legend lines in Figs. 10, 11, 12, 13, 14, 15 with changes in direction, indicates changes in layer properties, that is influenced by the layer thicknesses.

The processed data from AFCP-LVR software were presented graphically, showing the SPBC on horizontal axis and penetration depth on vertical axis. According to this information, a first attempt at layer interface recognition was made by considering changes in the slope of the graph. Figures 10, 11, 12, 13, 14, 15 showed 4 discerned structural layers, a medium strength wearing course to a depth of 100 mm, a strong base with a depth of 550 mm followed by a strong lower base to 1260 mm and sub-

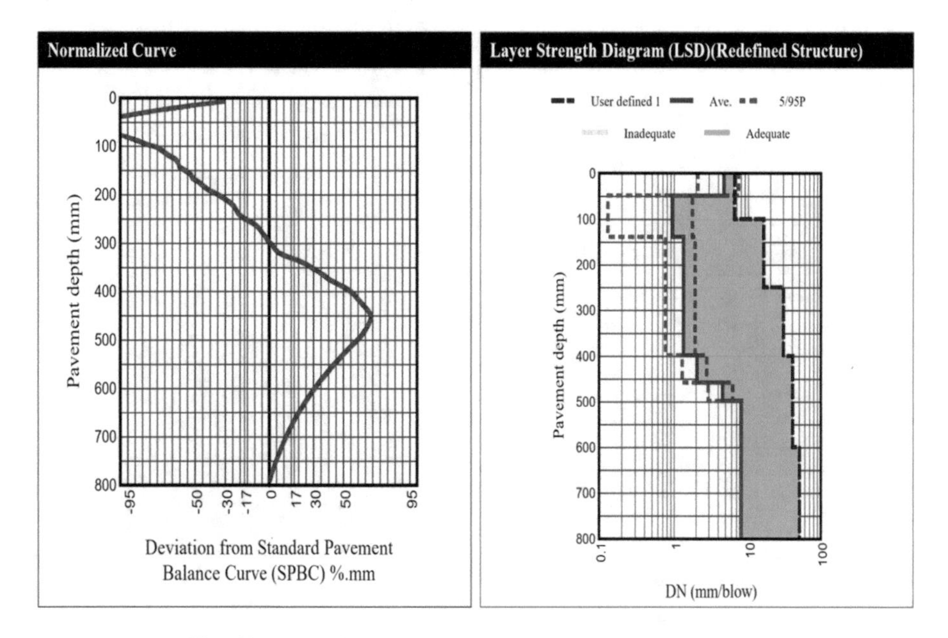

Fig. 10. Normalized curve and layer strength diagram Seg.1

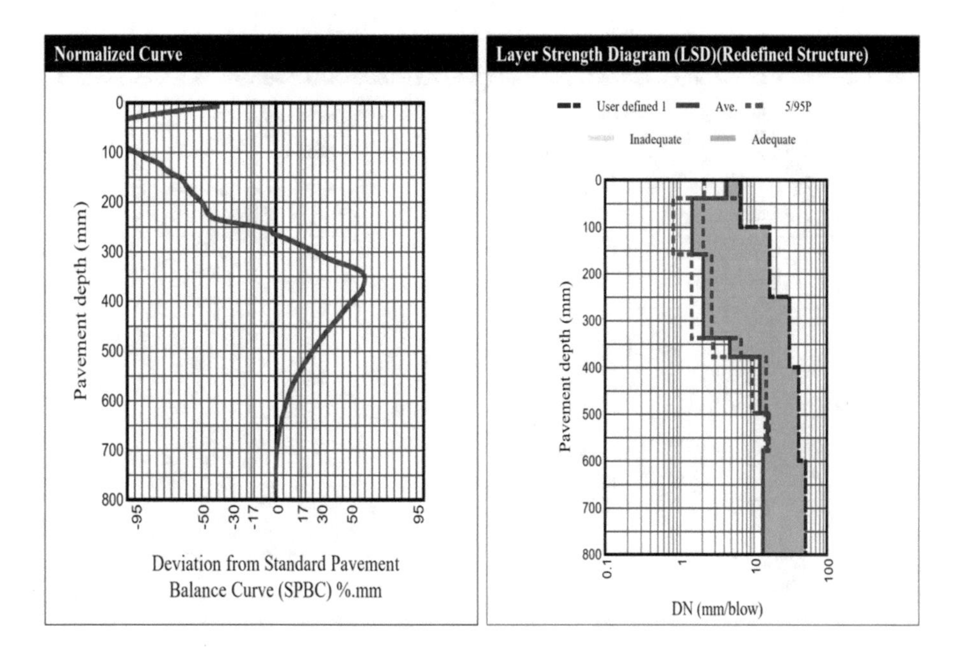

Fig. 11. Normalized curve and layer strength diagram Seg.2

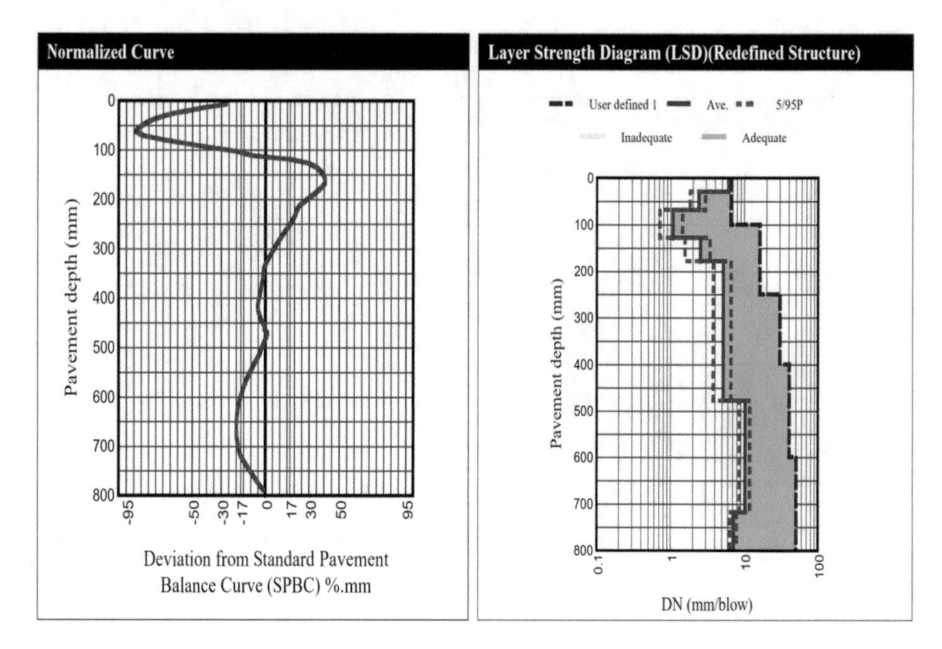

Fig. 12. Normalized curve and layer strength diagram Seg.3

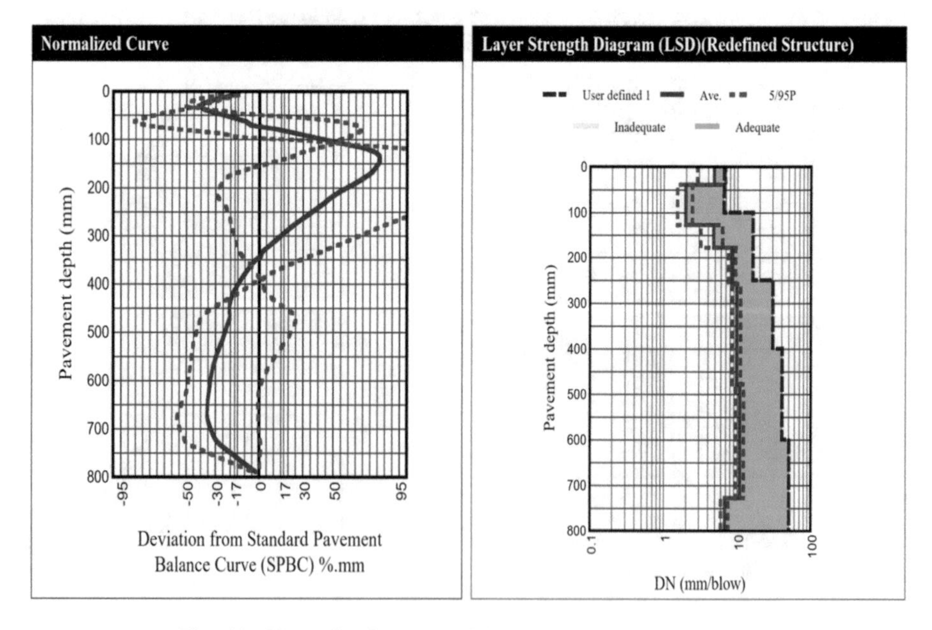

Fig. 13. Normalized curve and layer strength diagram Seg.4

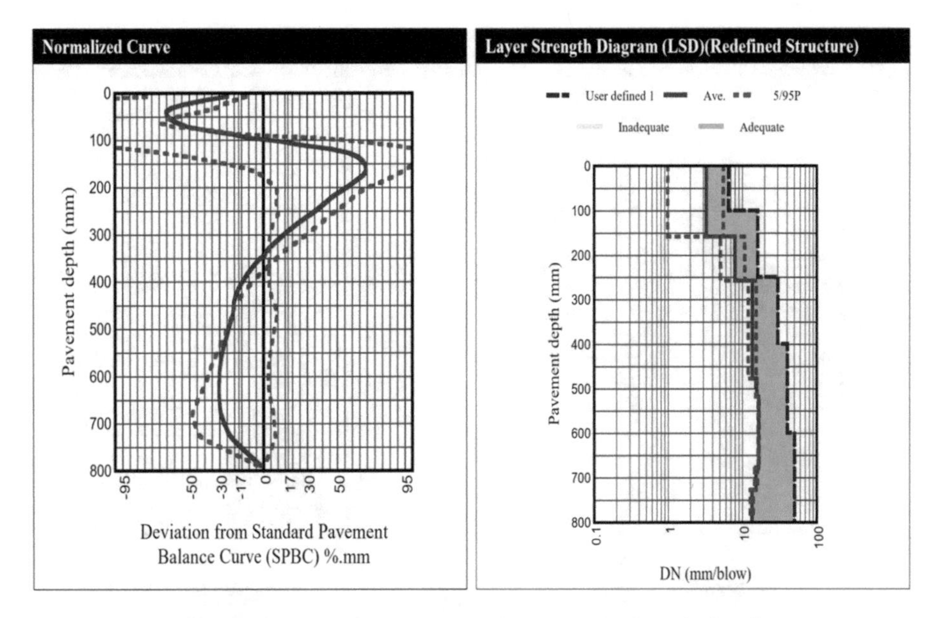

Fig. 14. Normalized curve and layer strength diagram Seg.5

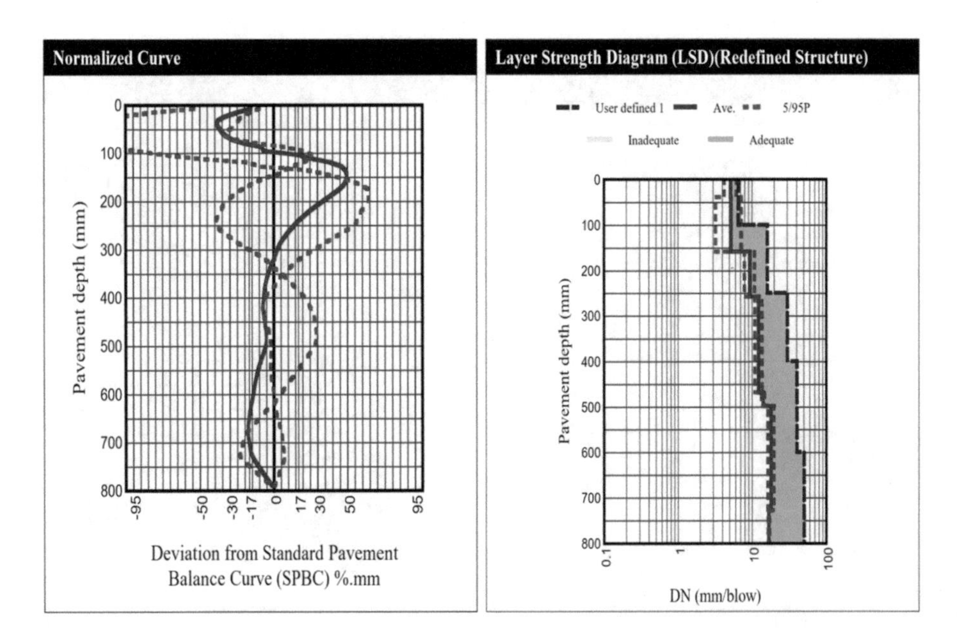

Fig. 15. Normalized curve and layer strength diagram Seg.6

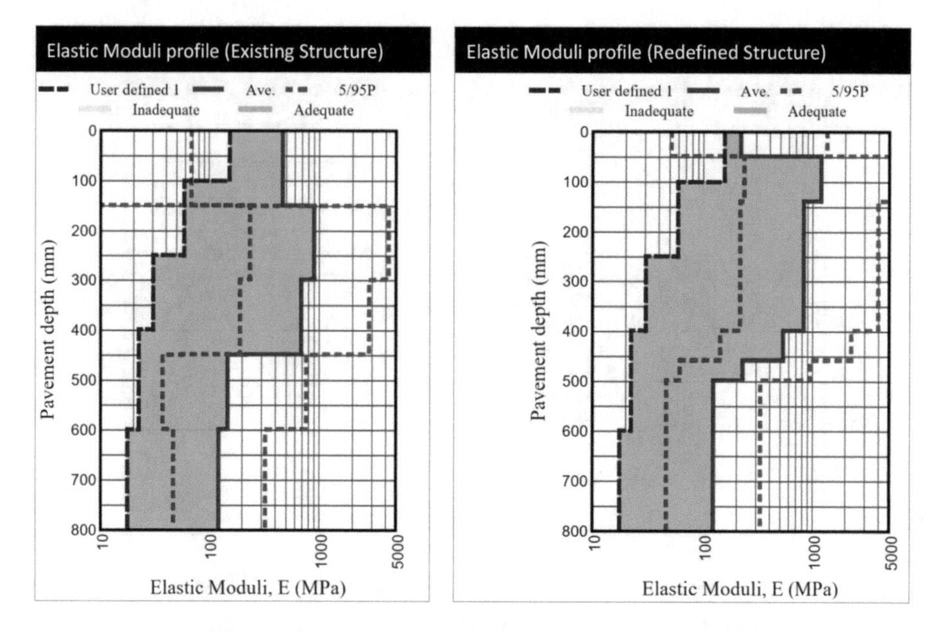

Fig. 16. Elastic moduli versus pavement depth Seg.1

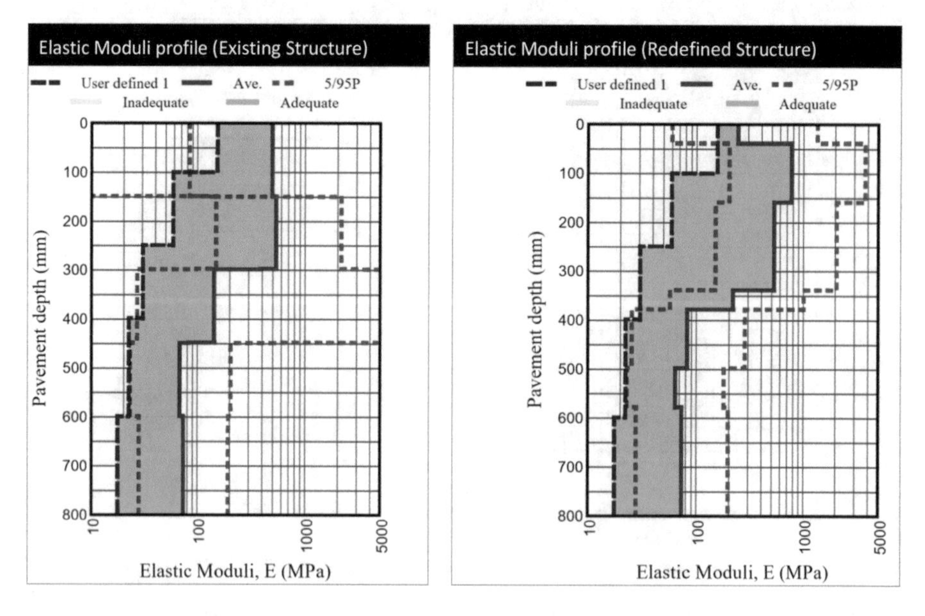

Fig. 17. Elastic moduli versus pavement depth Seg.2

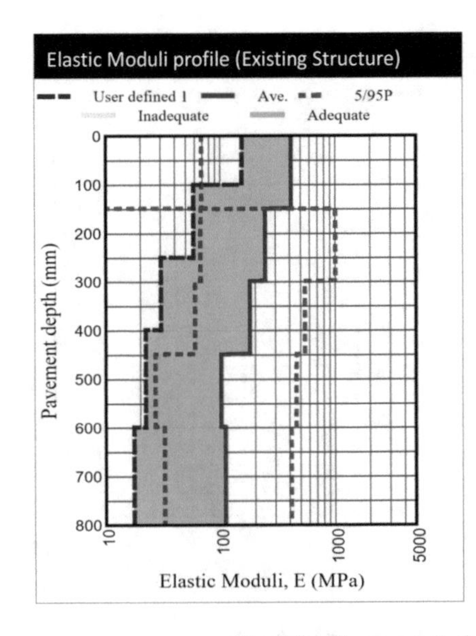

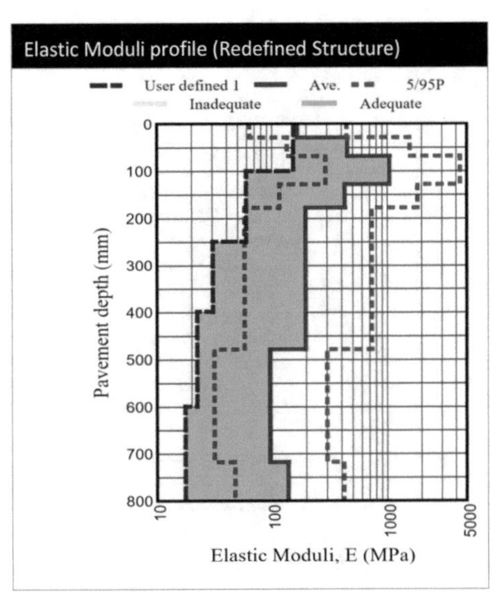

Fig. 18. Elastic moduli versus pavement depth Seg.3

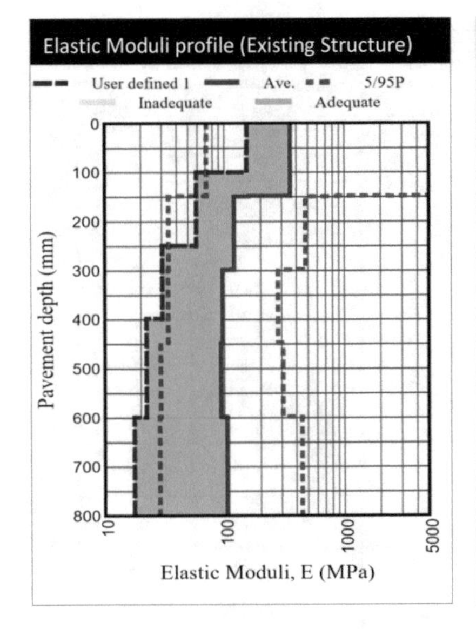

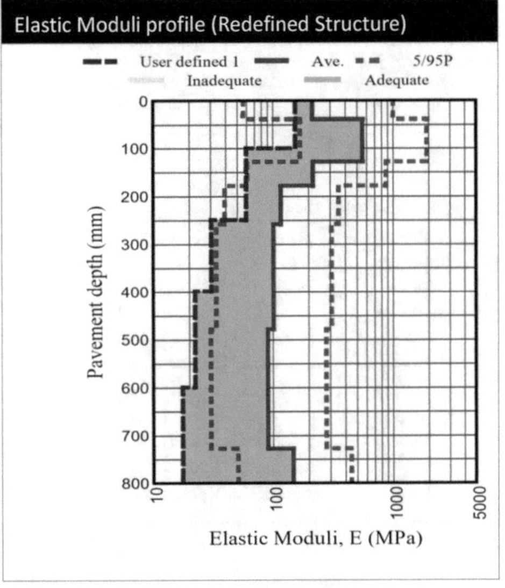

Fig. 19. Elastic moduli versus pavement depth Seg.4

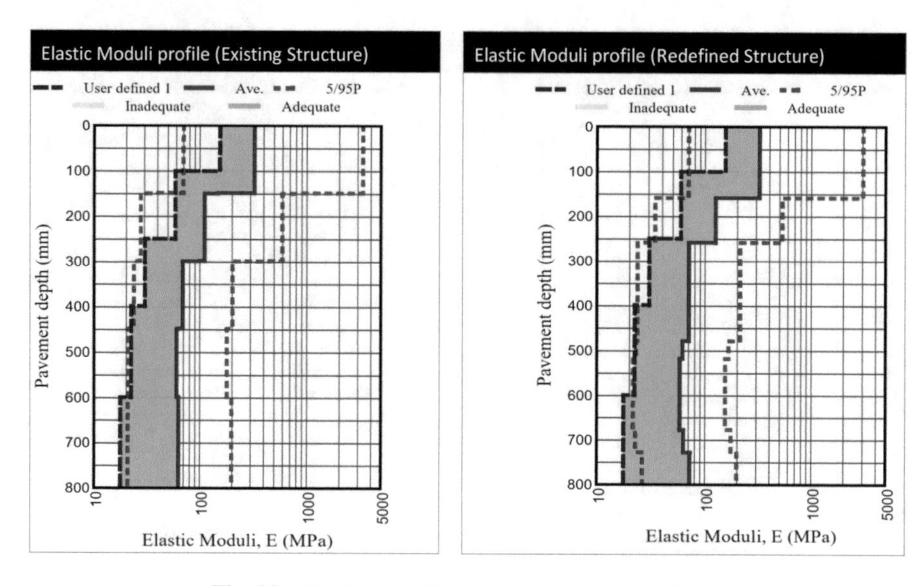

Fig. 20. Elastic moduli versus pavement depth Seg.5

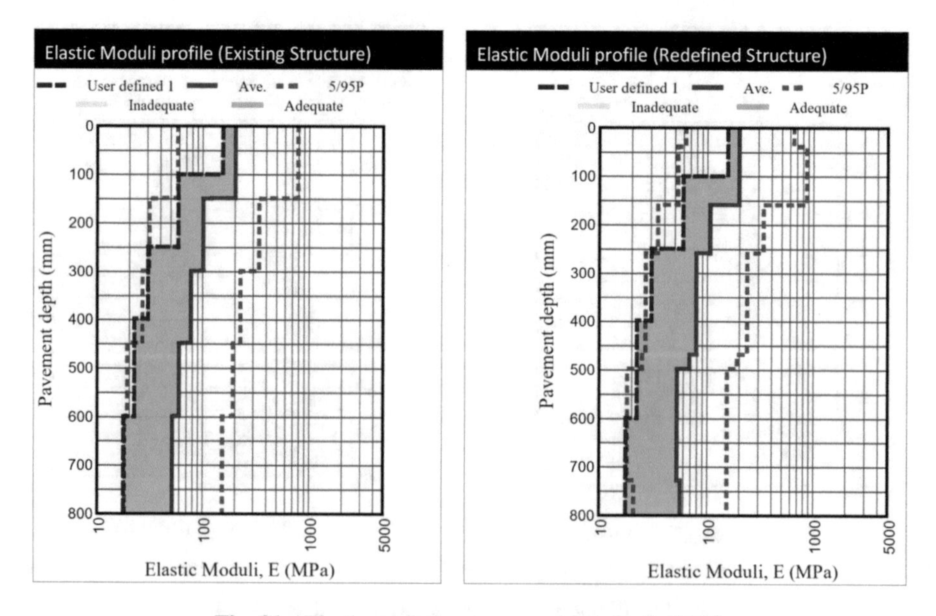

Fig. 21. Elastic moduli versus pavement depth Seg.6

base extends to the depth of 1600 mm. The ranges are recognized for SPBC in %. mm. When SPBC is greater than 40 the pavement is considered shallow pavements, between 0 to 40 it is designated as deep pavements and less than 0 is for inverted structures. The investigated pavement is confirmed to have a wearing course of 100 mm of a shallow pavement because the curve at each section cut-across beyond 40 mm Figs. 10, 11, 12, 13, 14, 15. Therefore, this confirmed the existing subgrade at each segment to be adequate for medium volume road traffic.

4.7 Elastic/Resilient Modulus Analysis

Three classes of pavements design expressed in Million Standard Axles (MISA) are: (a) Light traffic: less than 200 000 E80 s, (b) Medium traffic: between 200 000 and 800 000 E80 s, and (c) Heavy traffic: between 800 000 and 1.2 million E80 s (IRC:37-2001). Basically, to investigate required structural strength of the existing pavement, DCP field data expressed in terms of the DCP-layer-strength-diagram was projected to the appropriate DCP master curve as shown in Figs. 16, 17, 18, 19, 20, 21. The field data plotted on the left side of the design curves, signifies that the pavement foundation has adequate structural strength for medium traffic class. However, if the plot is located on right hand side of the design curve, this indicates that a region of the pavement structure with insufficient resilient strength.

In addition, DCP data curves plotted on the left side of a particular design curve, showing the area between the DCP field curve and the selected design curve is colored in green. This implies that adequate resilient strength is provided for the selected traffic class at a required depth. Furthermore, if DCP field data curve is located on the right side of a design curve, the area will be colored in yellow, signifying that the pavement structure at that depth does not possess enough structural strength to carry the selected traffic. Generally, there is no yellow legend lines on the curves and this implies that the existing pavement foundation is adequate for low to medium traffic loading.

Figures 16, 17, 18, 19, 20, 21 illustrates the distress mechanism of the existing pavement with respect to traffic loading. The field data of the existing pavement plotted in red, indicated that the pavement was designed not to carry more than 200 000 standard E80 traffic load applications while the Light traffic class is indicated as blue dotted lines (Figs. 15, 17, 18, 19, 20). The results in Figs. 15, 17, 18, 19, 20 illustrated the DCP field data for all the layers plot to the left of the selected design curve as presented by the green areas, which signify that the pavement has adequate strength to carry at least between 200 000 and 800 000 E80 s traffic load applications.

4.8 Traffic Count

Subsequently, during the study traffic count was conducted along the investigated pavement structure. The counts were conducted along Enugu – Abakaliki (CH 01 + 300) opposite Mobile filling station and Mbok – Ikom (CH 01 + 300) some meters before the big directional sign post. This was done in order to evaluate Average daily traffic (ADT) on daily basis along the pavement. However, the ADT value along the pavement section is 13,483 for Enugu-Abakaliki and this is the section where cracks/fatigues were discovered. Whereas, Mbok-Ikom recorded an ADT value of 846 and this investigated pavement section is still intact as a result of low volume flow of traffic. This result justified the analysis from the software, because the studied pavement was evaluated by the software are low-medium volume road.

4.9 Laboratory Resilient Modulus Results

Standard civil engineering testing procedures according to AASHTO T-307-99 were followed. Soils from each section were prepared using the field moisture content with specimens having size of 50 mm in diameter and 100 mm in height. The samples were tested with the combinations of three confining 20, 50 and 100 kPa at five deviatoric stress. M_r were determined by averaging the resilient deformation of the last five deviatoric cycles. Moreover, M_r increases with increasing confining pressure under constant deviator stress, which reflects a typical behavior of coarse material. This behavior is typical for the tested soils. The specimen subjected to higher confining stress yields higher M_r value and the effect of confining stress on M_r is relatively small as compared to that of deviator stress, which is expected for coarse soil. The results of the estimated M_r and CBR values compared with the laboratory evaluated values are presented in the Tables 3.

5 Conclusion

This paper evaluates the M_r response of an existing pavement, using DCP tools and software. In line with the objective, the study further examined and identified the possible causes of fast deterioration of this existing pavement structure along Eastern Nigeria (Enugu - Abakaliki – Ikom Highway Road) and following conclusions were reached:

The pavement sections were fatigue was discovered, were found to be superficially deformed as the failure is limited only to the asphaltic concrete layer. As the cracks has not extended beyond the wearing course.

Seg. 1, 2 and 4 fatigue and cracks were located can be resealed to prevent water percolation as to maintain good material resilient among pavement structure.

The laboratory CBR test was conducted in order validate the DCP predicted CBR values and AFCP-LVR software strongly predicted laboratory soaked CBR values despite the different testing procedures at a correlation value of $R^2 = 0.882$ on average.

DCP tools and AFCP-LVR software package is useful in evaluating pavement performance in terms of their balance, properties, layer thicknesses, etc. The software evaluates why some pavements performed well whilst performed poorly without rigorous laboratory exercise.

Acknowledgment. The Authors will like to acknowledge Geotechnical and Pavement Material laboratories as well as the academic staffs of Civil Engineering Department, Enugu State University of Science and Technology (ESUT).

References

AASHTO T 307-99: Determining the Resilient Modulus of Soils and Aggregate Materials, American Association of State Highway and Transportation Officials, Washington, D.C. (2003)

ASTM D 4959: Standard test method for determination of water (moisture) content of soil by direct heating. Designation D4959-00. West Conshohocken, PA (2000)

ASTM-D 6951-3: Standard Test Method for Use of the Dynamic Cone Penetrometer in Shallow Pavement Applications (2003)

ASTM D2937: Standard test method for density of soil in place by driven-cylinder method. American Society for Testing and Materials (2004)

ASTM D6913-04: Standard test method for particle size analysis of soil. American Society for Testing and Materials (2009a)

ASTM D4429-09: Standard test method for CBR (California Bearing Ratio) of Soils. American Society for Testing and Materials (2009b)

George, K.P., Uddin, W.: Subgrade characterization for highway pavement design, Final Report, Mississippi Department of Transportation, Jackson, MS (2000)

Gudishala, R.: Development of resilient modulus prediction models for base and subgrade pavement layers from in situ devices test results. Ph.D. thesis, Sri Krishnadevaraya University, Anantapur, India (2004)

Hassan, A.: The effect of material parameters on dynamic cone penetrometer results for fine-grained soils and granular materials. Ph.D. dissertation, Oklahoma State University, Stillwater, Okla (1996)

Herath, A., et al.: The use of dynamic cone penetrometer to predict resilient modulus of subgrade soils, Geo-Frontiers 2005, Geotechnical Special Publication ASCE, Reston (2005)

IRC (2002): Guidelines for the design of flexible pavements. Indian roads congress Number, vol. 37, New Delhi (2002)

Jianzhou, C., et. al.: Use of falling weight deflectometer and dynamic cone penetrometer in pavement evaluation, Paper Presented in the Transportation Research Board, Washington, D. C (1999)

Kleyn, E.G., et al.: The development of an equation for the strength balance of road pavement structures. Civ. Eng. S. Afr. **31**(2), 1989 (1989)

Li, D., Selig, E.T.: Resilient modulus for fine grained subgrade soils. J. Geotech. Eng. **120**(6) (1994)

Mohammad, L.N., et. al.: Prediction of resilient modulus of cohesive subgrade soils from dynamic cone penetrometer test parameters. J. Mater. Civ. Eng. **19**(11) 986–992 (2007)

Nguyen, B.T., Mohajerani, A.: A new lightweight dynamic cone penetrometer for laboratory and field application. J. News Aust. Geomech. Soc. **47**(2), 41–50 (2012)

Scala, A.J.: Simple methods of flexible pavement design using cone penetrometer. N. Z. Eng. **11** (2), 34–44 (1956)

Yoder, E.J., Witczak, M.W.: Principles of Pavement Design, 2nd edn. Wiley, New York (1975)

Phosphogypsum Management Challenges in Tunisia

Hajer Maazoun and Mounir Bouassida(✉)

Université de Tunis El Manar, École Nationale D'Ingénieurs de Tunis, LR14ES03, Ingénierie Géotechnique, BP 37 Le Belvédère 1002, Tunis, Tunisia
maazoun.hajer@gmail.com, mounir.bouassida@enit.utm.tn

Abstract. The accumulation of phosphogypsum (PG) produced till 2015 makes its management a real challenge to the Tunisian authorities and put the Chemical Tunisian Group (TCG) to face a challenge at large scale as the specified storage embankments knew considerable extensions in terms of heights and areas. Several studies were elaborated subsequently in 2007, 2012 and 2013 to focus on the stability of Sfax and Skhira Phosphogypsum embankments' and showed two different chemical and mechanical behaviors according to the experienced deposition process. In 2012, it was revealed that the wet PG embankment of Sfax City with 56 m height, 53Ha area and 32° slope can only be of 70 m height maximum. This embankment can reach 100 m in height if a reinforcement technique will be used. This deposition process is well recommended to ensure better interaction between the embankment and the existing ground surface. Using the dry deposited process, the area of the PG embankment of the Skhira City covers 112Ha and presents two elevation levels of 25 m and 55 m in 2013. However, the dry deposited process results in a damaged embankment profile, excessive settlements and lateral displacements. Therefore, a PG embankment of 100 m height cannot be achieved. A reinforcement of the embankment by High Density Polyethylene geotextile (HDPE) layers at increments of 4 m from 55 m elevation allows reaching 130 m of height. Comparative study was raised between the wet and the dry process and resulted in favoring of the wet process from both industrial and geotechnical perspective. Thus, the TCG expects turning all its deposition processes to the wet one.

1 Introduction

Phosphogypsum ($10(CaSO_4 2H_2O)$) is an industrial residue resulting from the phosphate ore attack by sulfuric acid when producing phosphoric acid (P_2O_5), of which 95% is used for chemical fertilizers and animal feed additives yearly. It is obtained according to the following chemical reaction Zairi and Rouis (1999), Felfoul et al. (2004):

$$\left[Ca_3(PO_4)_2\right]_3 CaF_2 + 10H_2SO_4 + 20H_2O \xrightarrow{70\ to\ 80\,^\circ C} 6H_3PO_4 + 10(CaSO_4 2H_2O) + 2HF$$

S. Hemeda and M. Bouassida (Eds.): GeoMEast 2018, SUCI, pp. 88–104, 2019.
https://doi.org/10.1007/978-3-030-01941-9_7

Nevertheless, this leads to many drawbacks. In fact, producing one tone of (P_2O_5) gives 5 tons of Phosphogypsum, leading to excessive quantities of PG which management is an environmental challenge.

Indeed, PG used to be deposited into the sea and oceans like in the American United States, Spain and the United Kingdom. However, this was forbidden by the 1990's for environmental concerns while the Morocco continued depositing more than 15 million tons of PG annually into the Atlantic Ocean IAEA (2013), International Maritime Organisation (1972). At the same period, a rehabilitation of Sfax City coasts, Tunisia, was occurred to repair the environmental damages caused by the NPK factory IAEA (2013).

Several researchers studied the use of PG in many fields, in an attempt to its use in different ways. However, the most successful valorization axis ever found is in fertilizing the saline soils, as experienced in Huelva, Spain Valverde-Palacios et al. (2011), Hilton (2010). Since the valorization possibilities in Tunisia are quiet limited, the management of the Phosphogypsum which is rising by 12 million tons yearly is a real challenge to the Tunisian authorities. Except in Gabes City, where it is still deposited into the Mediterranean Sea, the TCG is storing the phoshogypsum into embankments, known in Tunisia as "TABIAS", in the vicinity of the production units using the wet deposited process in Sfax and Mdhilla Cities and the dry deposited process in the Skhira City site TCG (2014).

Nowadays, the TCG faces a challenge of large scale as the specified embankments have known considerable extensions in terms of heights and areas. In fact, Tunisian wet phosphogypsum embankments heights have reached almost 56 m Bouassida (2012). For the wet embankment at Sfax site, it was predicted that the maximum height cannot exceed 100 m unless a reinforcement technique is used Bouassida (2012). Few similar cases are faced by countries involved in phosphate activity around the world as the wet phosphogypsum embankments heights do not exceed 28 m in Huelva, Spain Valverde-Palacios (2011), and are almost of 20 m in Mianzhu City, Baiyi Village, China www.greenpeace.org (2013), while the New Wales facility wet stack at Mulberry, Florida is expected to reach almost 91 m in 2023 www.reuters.com. Till June 2015, topographic updated land surveys showed that the dry phosphogypsum embankment of the Skhira City, comprising 112Ha, presented two levels of 25 m and 55 m successively with 78 m in its highest altitude. As the embankments increase in height, stability problems like cracks and slope displacements begin to appear in the dry embankment of Skhira. In 2012, it was proposed to cover the embankment by HDPE layers of 19MN/lm of modulus of rigidity at increments of 4 m of height since its top. This reinforcement allows elevating the embankment till 130 m of height Chaari (2013). Although, according to the TCG this solution not only necessitates the use of important quantities of geotextile layers but also requires excessive energy, developed equipments and numerous staff to flatten the damaged profile resulting from the deposition using dry process.

As phosphoric acid production is progressing, Phosphogypsum embankments are expected to receive important quantities of PG. The establishment of new deposition sites is a possible assumption typically in the Skhira City where it was meant to create a 40Ha surface and 42 m height wet deposited embankment. But, this site was expected to provide an exploitation period of 7 years only as reported the environmental impact study that launched the TCG in 2012 FNAC (2012). Hence, as the PG storage is ground intensive, it is time to optimize the storage sites and to search for another autonomous alternative: the recuperation of the 112Ha existing embankment deposited by dry process and its re-use as a support for a new embankment. No significant similar cases were found in the bibliography concerning dry stacks.

It was also pointed out that the major problems related to the dry process are typically geotechnical and environmental ones such as slope stability, settlements, friction angles and land wastes TCG (2014). According to its observations, the TCG mentioned that the PG deposition using wet process seems to be more beneficial than the dry process. Hence, it was particularly interesting to focus on the reliability of both of them via a justified scientific study, allowing this way the TCG authorities in the Skhira City to take the decision about conserving the actual deposit process or converting it into the wet one.

This work aims at presenting a synthesis of the different studies investigated to focus on the stability of Sfax and Skhira Phosphogypsum (PG) embankments as well as we introduce the actual challenges of PG storage.

2 Phosphogypsum Deposit Processes

The International Atomic Energy Agency (IAEA) estimated the quantity of Phosphogypsum produced worldwide till 2013 to be about 3 billion of tons IAEA (2013). Only 15% of this quantity is put into valorization, the other 85% is stored at embankments using either wet or dry process Moalla et al. (2017).

2.1 Wet Process

When deposing by wet procedure, Phosphogypsum lefts the filter as sludge composed of almost 30% of solid particles and 70% of water. This sludge is transported to the storage embankment via pipes and is evacuated in the TABIA's top using an evacuator (Fig. 1).

Fig. 1. Pipes reversing Phosphogypsum sludge via wet process – Sfax Unit

The embankment contains two cells separated by an intermediate dike resulting in two sedimentation basins exploited reversely. An exploitation cycle consists in filling in a basin while the other is getting re-managed.

The gypsum water is discharged in the highest extremity of the sedimentation basin. This way, the sludge is deposited under an area of 0.3 to 0.4% of inclination to the sea. Hence, water moves through gravity, gets evacuated via a valve and reaches a recuperation basin. This drained water is then pumped to the factory for re-use and almost only 20% of the discharged water is wasted by evaporation and infiltration TCG (2014).

2.2 Dry Process

In dry process, Phosphogypsum leaves the filter as a powder (solid particles) of 30–35% of humidity. Then, it is transported to the deposit area by belt conveyors and deposited in the top of the TABIA (Fig. 2). The PG discharge is processed by progressing over the storage area and adding supplementary belt conveyors. Hence, dry process requests large deposition areas and results in damaged reliefs which require important equipments and labor for both of maintenance and management.

Fig. 2. Belt conveyors transporting Phosphogypsum to the deposit site via dry process – The Skhira City Unit

3 Study of Sfax Phosphogypsum Embankment Deposited Using Wet Process

In 2012, the PG deposit of Sfax which covers 53Ha was almost 56 m of height and 32° of slope. A study carried out by the National Engineering School of Tunis (ENIT) was led to focus on the embankment stability and identify the maximum height that can be reached without getting damaged. The study was based on the geotechnical site investigation consisting in realizing a destructive hole (SD) of 55 m of depth, 3 pressure meter holes (SP_1, SP_2 and SP_3) of respective depths: 60 m, 42 m and 42 m, 2 core tests (SC_1 and SC_2) of 10 m of depth each, Le franc and SPT tests and extraction of samples from different depths to be identified in the laboratory Bouassida (2012).

3.1 Geotechnical Aspect of Sfax Embankment

The reported investigation permitted to identify the geotechnical profile of both of the embankment and its ground surface.

In fact, according to the holes SD and SP_1, the embankment presents two different horizons; an upper layer of 8 m of thickness, with a cohesion evaluated at 10 kPa and a friction angle of 32°, while the lower layer (relying on land) cohesion is 4 times higher than the upper one. The pressure meter tests carried out as 1 test per 2 m show that the pressure meter modulus of the lower layer is 3 times higher than the upper layer one. The results are illustrated in Table 1.

Table 1. Mechanical characteristics of Sfax phosphogypsum embankment Bouassida (2012)

Layer	Thickness (m)	C (kPa)	(°)	$E_{m\ average}$ (MPa)	$Pl^*_{average}$ (MPa)	N_c
Upper	8	10	32	49	2.31	15
Lower (relying on land)	48	41.2	32.3	124.2	3.98	21

N_c: Number of corrected blows

As for the ground surface, it can be divided into 4 layers for which pressure meter modulus ranges from 9.9 MPa in surface to 143.2 MPa in depth (Table 2). No detailed geotechnical property of the ground surface of Sfax plant are available except of $E_{m\ average}$ and $Pl^*_{average}$. The ground surface presents excellent mechanical characteristics due to the compaction and the void index reduction caused by the accumulation of the embankment weight. This way, the study is led in the hypothesis that it is assimilated to a stratum.

Table 2. Mechanical characteristics of the ground surface Sfax phosphogypsum embankment Bouassida (2012)

Layer N°	Thickness (m)	$E_{m\ average}$ (MPa)	$Pl^*_{average}$ (MPa)
1	8	9.9	1.17
2	8	24.6	1.89
3	8	16.6	1.55
4	16	143.2	4.04

3.2 Grain Size Analysis and Settlement Estimation

The granulometric analysis showed that phosphogypsum is a coarse-grained to fine soil with a significant fine percentage when going in depth. Two main reasons are behind this particle size distribution; the first one is the deformation of the grains themselves under the strengths they apply on their contact points, the second is the reduction of the void index by re-messes of grains while the embankment elevation is progressing. Terzaghi and Peck correlation (1948) was used to determine the embankment settlement while supposing that it was constructed under respectively for steps: 3 layers of 16 m and one of 8 m thickness. After correction, the settlement was evaluated at 0.86 m Bouassida (2012).

3.3 Estimation of Maximum Embankment Height

The highest level that can reach the phosphogypsum embankment of Sfax was estimated using Flac 6.0 software and two main conclusions were drawn: The mechanical characteristics of the upper layer ($\varphi = 32°$, c = 10 kPa) allow going on height by 10 m yet. A height of 100 m can be reached if the embankment gets reinforced by geotextile layers, a technique that was actually the objective of a recent study carried out in 2013 concerning the dry depositing process Bouassida (2012).

4 Studies of Skhira Phosphogypsum Embankment Deposited Using Dry Process

The dry embankment of Skhira knew considerable extensions in terms of surface and height. Nowadays, it covers an area of 112 Ha in the basis and over time, it becomes suffering from much pathology threatening its stability. Hence, two different studies were carried out to focus on this respectively at 2007 and 2013.

4.1 Geotechnical Aspect of Skhira Embankment

Up to 2007, the PG deposit of Skhira presents two different levels of almost 25 m and 55 m of height and a slope varying between 1/4 and 2/3. The geotechnical profile of the ground surface reveals 4 horizons which characteristics are shown in Table 3. There is no problem in relation with the ground surface bearing capacity as a Factor of Safety of 2 is obtained for an elevation of 100 m of height, a unit weight γ of $19.7 kN/m^3$, an undrained cohesion and friction angle of $c_u = 85$ kPa and $\varphi_u = 31°$.

The embankment slope stability was verified by TALREN software (version 4 v 3.1) using BISHOP Theory. This allowed following the Factor of Safety while varying the embankment height and slope and revealed that the embankment can reach 60 m of height with a slope of 2/3 and a Factor of Safety F = 1.3 and can be elevated till 70 m with a slope of ½ URIG (2007).

Table 3. Geotechnical profile of the ground surface under the Skhira phosphogypsum embankment URIG (2007)

Layer	Thickness (m)	ω (%)	γ (kN/m^3)	C_u (kPa)	C_c	e_0	φ_{uu}
Gypsums silt	18	20	19.7	345	0.07	0.62	34°
Clay loam	2	20	19.6	345	0.07	0.62	34°
Sand	5	–	–	–	–	–	–
Clay loam	5	23	19.6	90	0.21	0.66	35°

4.2 Settlement Estimation

The ground surface and the embankment settlements were estimated using Terzaghi Theory and Costet and Sanglerat formula (1983) and resulted on a total settlement varing from 4.72 m to 4.91 m (Table 4).

Table 4. Settlement estimation of the Skhira embankment and ground surface URIG (2007)

Height (m)			10	20	30	40	50
Settlement (m)	Embankment		0.56	1.12	1.68	2.24	2.8
	Ground surface	SCT2	1.12	1.42	1.63	1.79	1.92
		SCT3	1.19	1.53	1.77	1.95	2.11

Hence, important settlements are predicted, caused essentially by the embankment's weight. Nevertheless, this can be reduced by reinforcing the embankment by geotextile layers; proposition that was treated in an anterior study in 2013 for embankments deposited using dry process URIG (2007).

4.3 The Maximum Reachable Embankment High Estimation

A study was carried out in 2013 using Plaxis numeric modelisation in order to determine the maximum height that can reach the Skhira dry Phosphogypsum embankment once it is reinforced by HDPE layers from different levels Chaari (2013). Mohr-Coulomb's failure criterion was adopted for the numerical modeling. No available justifications of the methodology for Slope stability analysis, parameters chosen and the sensitivity of the parameters are available. This part is missing in the study of report by Chaari (2013).

It supposes that the Phosphogypsum embankment is resting on a ground surface composed of 4 layers and that the embankment is 140 m height. The deposit is modelled as a superposition of 3 respective layers from the top-down: PG(1), PG(2) and PG(3) in such a way that each one presents different mechanical characteristics inspired from previous studies (Fig. 3).

The TCG experience shows that each 360 days, the Phosphogypsum embankment height increases by 4 m. From a loading stage (n) to another (n + 1), an amelioration of the mechanical characteristics of the layer (n) happens. From (n + 1) to (n + 2), the mechanical characteristics of the layer (n + 1) get improved, those of the layer

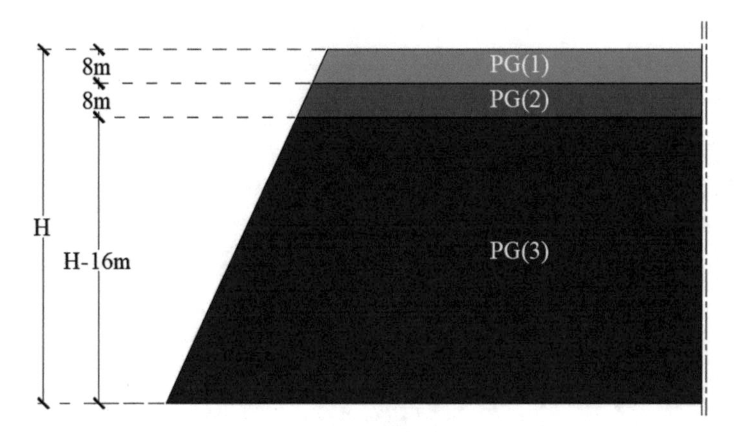

Fig. 3. The three layers PG(i) of the dry deposited embankment of the Skhira City Chaari (2013)

(n) increase also and over time, the layer (n) get integrated into the layer (n − 1) Chaari (2013). In what follows, the study supposes a fixed height of 8 m for each of the layers PG(1) and PG(2) and a variable height increasing at increments of 8 m from a design sequence to another where a design sequence corresponds to a loading phase. Three cases were treated: the first one is when the embankment is not reinforced. The second one is when it is reinforced from 55 m of height. The third one is when it is reinforced from its basis.

4.3.1 Unreinforced Embankment (Case 1)

Plaxis simulation showed that a 130 m height under a slope of 2/3 can be reached without instability risks for a Factor of Safety of $F_1 = 1.4$ (Fig. 4a). The resulting settlement of almost 2.6 m is expected with a maximum horizontal displacement in the embankment basis at a rate of 0.37 m (Fig. 5a).

Compared to the 2007 study, it is clear that due to the amelioration of the Factor of Safety by 0.1, the embankment height gets almost doubled. This gap was explicated by the manual design cumulated errors and the high Factor of Safety demanded by Plaxis software, although similar results were obtained using GeoStudio Slope software Chaari (2013).

4.3.2 Embankment Reinforced from 55 M of Height (Case 2)

The incorporation of High Density Poly Ethylene (HDPE) layers of a rigidity module of 19MN/lm at increments of 4 m from 55 m of height is expected to ameliorate the reported Factor of Safety by 35% ($F_2 = 1.9$) (Fig. 4b). Although, this reinforcement has no significant effect neither on the settlements nor on the horizontal displacement compared to the non reinforced embankment case (Fig. 5a and b).

4.3.3 Embankment Reinforced from Its Basis (Case 3)

In order to characterize the layers behavior as well as their eventual contribution to the settlement and horizontal displacements reduction, a third case was mentioned. The

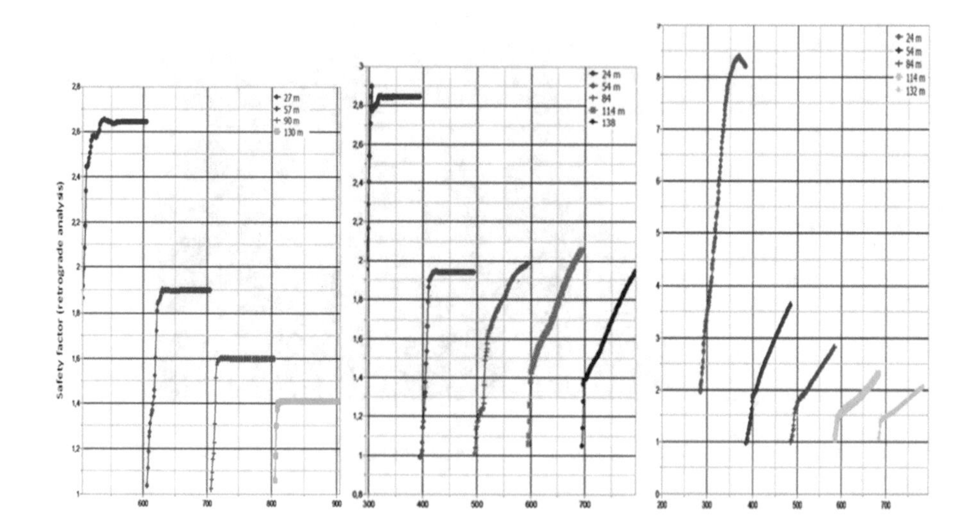

a. Unreinforced embankment
b. Embankment reinforced by HDPE layers from 55m of height
c. Embankment reinforced by HDPE layers from its basis

Fig. 4. Factor of safety vs. embankment height variations, Case study 1, 2 and 3, respectively Chaari (2013)

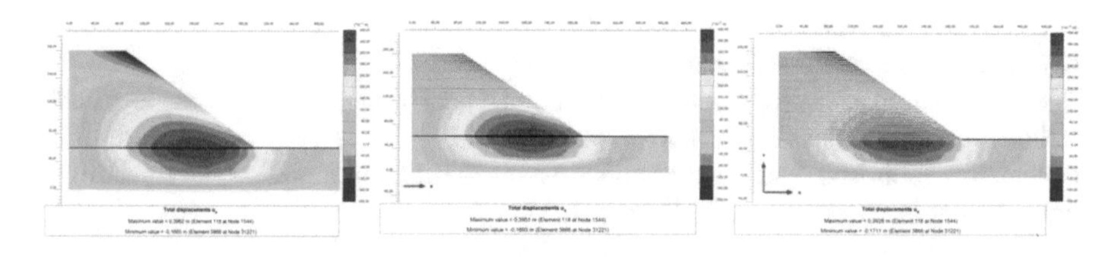

a. Unreinforced embankment
b. Embankment reinforced by HDPE layers from 55m of height
c. Embankment reinforced by HDPE layers from its basis

Fig. 5. Displacements vs. embankment height variations Case study 1, 2 and 3, respectively Chaari (2013)

studied model is identical to the second case with reinforcement by HDPE layers of identical characteristics from the embankment basis by 4 m increments. The results given by Plaxis software provided a Factor of Safety of only $F_3 = 2.06$ which is insignificant compared to $F_2 = 1.9$ (Fig. 4b and c).

However, this confirms the hypothesis that Phosphogypsum mechanical characteristics increase by time and that the PG (3) layer's Young module which is of 194 MPa is reliable to insure the embankment stability. Break risks are rather probable in the superior 16 m of the deposit (PG (1)), in whatever construction level of the embankment (Fig. 3).

From a mechanical concern, this simulation proved an excellent adherence in the interface embankment-HDPE illustrated by an equal displacement of both of them. No significant amelioration was mentioned besides that.

4.4 Study Main Outcomes' and Limits'

It could be concluded that the embankment reinforcement using HDPE layers has not any effects on the settlement and the lateral displacement and remains insignificant in the bottom of the deposit where Phosphogypsum already acquire the mechanical characteristics of a stratum. The incorporation of HDPE layers from 55 m in height provides higher Factor of Safety which allows very high embankment.

Nevertheless, this reinforcement using HDPE is limited by three main factors: First, the Tunisian Phosphogypsum pH is found to be of 2.9 Ajam et al. (2009) and about 3 Felfoul et al. (2002a). This high acidity of Phosphogypsum affects the integrity of HDPE layers. Although, it was proved that it increases by time as it is of 2.63 outlet filter, 2.96 when Phosphogypsum is aged 10 years and 3.24 when has been stored for 50 years Felfoul et al. (2003). Second, because of the accumulation of Phosphogypsum by storing, the thickness of layer PG(3) is in a continuous increase. Each Phosphogypsum discharging cycle defines a new thickness of this layer and the layers PG(1) and PG(2) (the upper 16 m of the embankment) are always active. Over time, there will be an accumulation of HDPE layers partially included into the layer PG(3) which is assimilated to a stratum and does not require any reinforcement (Fig. 3). Third, the study did not take into consideration the observed cracks within the embankment as reported by TCG engineers. So, these results cannot be achieved in reality unless the deposit procedure might be improved.

5 Performance Monitoring of the Embankment for the Start of Embankments' Construction to the Current Stage

The variation range of Tunisian Phosphogypsum chemical composition as presented by the TCG in 1995 is shown in Table 5 Belaiba et al. (2004). By 2017, The Skhira plant laboratory provided a recent chemical analysis of the dry deposited Phosphogypsum composition and showed slight different results (Table 5). In fact, the chemical composition of Phosphogypsum depends on the origin of the phosphate ore, the manufacturing process, the efficiency of the plant and the age of the deposit Choura et al. (2015). These characteristics evolve over time like the soluble P_2O_5 which content increases as the PG gets older due to rain wash for example Felfoul et al. (2003).

The nature and characteristics of the resulting phosphogypsum are strongly influenced by the phosphate ore composition and quality Sahu et al. (2014). Hence, Phosphogypsum is considered to be radioactive as it derives from the naturally radioactive phosphate ore. This radioactivity is due to the radium content coming from the natural decomposition of uranium. The USEPA has classified Phosphogypsum and rock phosphate as "Technologically Enhanced Naturally Occurring Radioactive

Table 5. Tunisian Phosphogypsum chemical composition

Element	Content (1995) - Tunisian PG	Content (2017) – Tunisian dry deposited PG
P_2O_5	0.063 – 0.197%	1.5%
CaO	31.9 – 32.14%	32%
SO_3	44.58 – 44.75%	44%
SiO_2	1.73 – 2.27%	1.8%
Al_2O_3	0.13 – 0.16%	0.1%
Fe_2O_3	0.09 – 0.10%	0.05%
MgO	0.01 – 0.02%	0.01%
Na_2O	0.12 – 0.16%	
K_2O	0-0.01%	
F	0.6 – 1.2%	1.5%
Cd	23–35 ppm	
Organic C	0.33 – 0.64%	0.8%
Water content	20–35%	32%

%: Percentage by weight

Materials" (TENORM) and Phosphogypsum exceeding 370 Bq/kg of radioactivity has been banned from all uses by the EPA since 1992 Sahu et al. (2014). For the Tunisian Phosphogypsum, The ^{238}U and ^{232}Th activities (47 Bq/kg and 15 Bq/kg) are as low as the average concentrations of these radioelements found in Tunisian soils and thus do not present any risk for the environment. The activity of the ^{226}Ra found in Tunisian PG (215 Bq/kg) remains lower than those found for the majority of PGs Ajam et al. (2009).

The mechanical characteristics of the wet deposited Phosphogypsum over time were studied basing on three specimens: SP1: outlet filter, SP2: aged 10 years and SP3: aged 50 years Felfoul et al. (2003). The specimens SP2 and SP3 are superficial. The study proved that the best bearing capacity is obtained for SP2 (CBR = 51% compared to 49% and 5% for SP3 and SP1 respectively) as well as for the best shear strength ($c = 73$ kPa and $\varphi = 37°$). The greatest part of the compressive and the tensile strengths are developed by the 15 first days of Phosphogypsum deposition with a better mechanical performance and behavior to water for the specimen SP2. The study is based on the strengths ratio defined as the ratio of the strength (compressive or tensile) after immersion to the strength without immersion. This ratio is null immediately after immersion and stills so for SP1 even 120 days after immersion. This indicates a very bad behavior to water of the fresh Phosphogypsum. For SP2, the strengths ratio gets improved for 1.13 to 1.36 times 24 h after immersion and doubles after 5 days. This indicates that as a result to the bad weather, the strengths fall but can get improved by time. This good behavior to water is explained by the high cohesion of the Phosphogypsum at a young age (SP2: $c = 73$ kPa, SP1: $c = 53$ kPa and SP3: $c = 50$ kPa) Felfoul et al. (2003). For the specimen SP3, the strengths go through an optimum at 28 days of the samples preparation. However, the strengths ratio decreases with the age.

The fresh Phosphogypsum wash allowed improving the compressive strength which reveals that the decrease of acid and organic contents enhances the compressive strength of Phosphogypsum. Hence, the chemical evolution of Phosphogypsum is behind its mechanical characteristics variation rather than its age Felfoul et al. (2003).

Performance monitoring of the dry embankment are not mentioned in the bibliography, there are only observations deduced from field visits carried out in March 2017. In fact, when deposited under the conventional water content (30–32%), a clear heap form is obtained. The Phosphogypsum starts to dry and local subsidences acquire at the foot of the heap Chaari (2013). The deposition of Phosphogypsum with a water content $\omega > 30–32\%$ results in slides at the top of the deposit assimilated to mud-flows. The material behaves like a liquid and the flows increase as the deposited quantity of Phosphogypsum increases. Undulations reaching the foot of the embankment are observed instead of the heaps. One day later, water seepage and horizontal cracks are obtained and there are shifts of dried out areas (Fig. 6).

Fig. 6. Phosphogypsum deposited with a water content $\omega > 30–32\%$ URIG (2007), Chaari (2013) and Bouassida (2012)

The deposition of Phosphogypsum at a high-water content ($\omega \ggg 35\%$) resulted in a big puddle of water at the top of the embankment (deposition line C) where the Phosphogypsum particles got settled. Important clear water seepage was observed by this deposition line.

A similar phenomenon occurred in the deposition line B where there was deposition of Phosphogypsum with very high-water content according to the CTG. The water was so abundant that it was impossible to install a belt conveyer there as it does sink into the ground. Hence, the deposition line was redirected and a drainage ditch of 1 m of depth was digged among the area. The water seepage has been observed for more than 4 months which proves the presence of water retained into the embankment body. In a larger scale, experience with dry storing indicates that the lower layer of the embankment will be saturated even in desert, in arid climate and without rain infiltration due to gypsum self-weight consolidation and settlement Fuleihan (2012). This occurs to some extent in all dry stacks even when Phosphogypsum is well filtered ($w \ll 25\%$).

6 Comparative Study Between the Wet Process and the Dry Process Deposition

A comparative study is launched to focus on the reliability of the wet and the dry deposition processes. The study concerns 6 successive years of phosphoric acid (P_2P_5) production ranging from 2005 to 2010 and focuses on the parameters affecting the cost of PG storage.

6.1 Parameters and Data Used

The existing embankments in Sfax and the Skhira Cities areas and contents are shown in Table 6. In what follows, "X" designs the dry discharge of the Skhira City and "Y" the quantity of stored Phosphogypsum in the same area till 2010. Hence, in an area of almost 0.5X Ha, can be stored a PG quantity of 0.88Y Tons using wet process. This way, an area of X Ha can receive 1.76Y Tons of PG if proceeding by the wet method, which means a win of 76% on the stored quantity.

Table 6. Sfax and the Skhira Phsphogypsum embankments areas and contents

	Wet process	Dry process
	SFAX	SKHIRA
Discharge area (Ha)	53	112
Stored PG quantity till 2010 (T)	36 798 300	42 021 000
Stored PG quantity by m^2 (T/m^2)	69,431	37,519

Wet process has always been known as less safe to land and ground water as almost 64% of dumped water is either evaporated or infiltrated in the basis soil TCG (2014). Although, due to the use of geomembrane, an important portion of the evacuated water can be drained and reused in the industrial process. Like so, even if the water consumption in the wet process is almost 6 times the dry one (Table 7), this seems to be less dangerous than the dust emission which is inevitable in the dry process and which causes air pollution and disturbs habitations near the production site. The water recovery does not only save the environment, but also has an effect on the industrial efficiency: According to the last statistics given by the TCG's phosphoric acid production units, this effect in the wet process ranges from 0.5% to 1% while it is of 0.42% only in dry process. This makes clear that the water losses are particularly important in dry process, which not only reduces the industrial efficiency, but also the water quantity returned for reuse in the factory.

Table 7. Water balance for the wet and the dry depositing processes TCG (2014)

$m^3/T(P2O5)_{produced}$	Wet process	Dry process
Water consumption	13.5	2.4
Recovered water percentage (%)	31%–36%	25%

The previous studies focusing on the stability of the embankment of the Skhira City show some limits of the dry process of a geotechnical concern (Table 8). In fact, according to the 2007' study, the embankment settlements due to its own weight were estimated about 2.8 m, resulting in a total settlement varying between 4.72 m and 4.91 m in the tested points against a maximum of 0.86 m only in the wet embankment of Sfax Bouassida (2012).

Table 8. The dry and the wet embankments properties

<table>
<tr><th>Embankment deposited by</th><th colspan="3">Dry process URIG (2007) and Chaari (2013)</th><th colspan="3">Wet process Bouassida (2012)</th></tr>
<tr><td>Area (Ha)</td><td colspan="3">53</td><td colspan="3">112</td></tr>
<tr><td>Height (m)</td><td colspan="3">56</td><td colspan="3">2 levels: 25 and 55</td></tr>
<tr><td>Mechanical characteristics</td><td>Height (m)</td><td>c (kPa)</td><td>φ (°)</td><td>Height (m)</td><td>c (kPa)</td><td>φ (°)</td></tr>
<tr><td>– Layer 1(upper)</td><td>8</td><td>90</td><td>30</td><td>8</td><td>10</td><td>32</td></tr>
<tr><td>– Layer 2</td><td>8</td><td>97.2</td><td>30</td><td>48</td><td>41.2</td><td>32.3</td></tr>
<tr><td>– Layer 3</td><td>$H^{(*)}$–16</td><td>104.4</td><td>30</td><td colspan="3" rowspan="4"></td></tr>
<tr><td></td><td colspan="3">Chaari (2013)</td></tr>
<tr><td>The hole embankment</td><td>H</td><td>85</td><td>31°</td></tr>
<tr><td></td><td colspan="3">URIG (2007)</td></tr>
<tr><td>Total settlements (m)</td><td colspan="3">4.72–4.91</td><td colspan="3">0.86</td></tr>
<tr><td>Lat. displacement (m)</td><td colspan="3">≈0.4</td><td colspan="3">Insignificant</td></tr>
<tr><td>General aspect</td><td colspan="3">– More damaging settlements and displacements apart from the cracks and the slope instability</td><td colspan="3">– Regular top and slope
– Better mechanical behavior of the embankment and the ground surface</td></tr>
<tr><td colspan="7">$H^{(*)}$: the height of embankment at an arbitrary level of its deposition</td></tr>
</table>

The Tunisian Phosphogypsum embankments are not instrumented. Hence, no actual measurements are available to follow up the settlement measurements

The settlement of an instrumented experimental Belgium Phosphogypsum embankment was followed-up during 30 months Gorlé and Reichert (1985). The embankment is 5.2 m of height and 50000 m^3 of volume and is executed by compaction of consecutive layers of Phosphogypsum of 0.2 m of thickness. After two years follow-up, the embankment settlement is still progressing while the settlement of the ground surface and the sand layer underlying the embankment is maximal after 6 months. The extrapolation of the results over a ten-year period at linear and logarithmic scales gives 0.2 m and 0.06 m respectively Gorlé and Reichert (1985) and 1.17 m over 61 years. Predictions using Terzaghi et Peck correlation (1948) indicate the wet deposited embankment of Sfax City should settle by 0.86 m since its construction to current time (1952–2012) Bouassida (2012). The difference between the experimental and the numerical results can be explained by:

- The instrumented embankment is executed through Phosphogypsum compaction; it does not result from the conventional Phosphogypsum deposit processes.
- The results extrapolation does not take into consideration that over time, the settlement slows down.

Besides, both of the two studies do not take into consideration the deposit cycles.

A better stability was obtained for the wet embankment with a slope of 2/3 and with a Factor of Safety of 1.02 in opposition to 1/4 in some sensible parts of the dry one under a factor of 1.3, Bouassida (2012).

Hence, the mechanical characteristics of phosphogypsum are best performing in the wet process than the dry one, it gives better results with a regular top and slopes and provides better mechanical behavior of the embankment and the basis soil than the dry one. The settlements and displacements are more damaging in the dry process, apart from the cracks and the slopes instability problems.

The TCG experience shows that the stops rate by breakdowns is of just 0.01% for the wet process while it is of 3.83% for the dry one. As a consequence, the phosphogypsum embankment of the Skhira City needs a lot of care and reparation, its maintenance cost is higher than the wet deposited embankment case TCG (2014).

6.2 Main Results

Based on the reported reasons and on the TCG staff's know-how and experience, it is clear that the wet process is more advantageous than the dry one. In fact, depositing phosphogyosum using wet process allows stacking almost 76% of PG more than when using dry method. In addition, although the wet process is 6 times more water consumer, it allows a better recuperation of soluble P_2O_5, which is translated by an effect on the industrial efficiency ranging from 0.5% to 1%. Besides, there is almost no shutdowns caused by the circuit breakdowns in wet process, the production is only stopped for the planned maintenance periods and it mobilizes a fewer staff. As a consequence, the maintenance cost in wet process is around 20% of the cost in dry process.

From a geotechnical concern, a better stability is obtained for the embankments deposited by wet process with better mechanical characteristics (cohesion and friction angle), less settlements and horizontal displacements as well as more regular slopes and top.

For all these reasons, the TCG is oriented to adopt from now on the wet process to stack its residues and is planning to modify its equipment in the Skhira factory and replace it by a wet process circuit.

7 Conclusions and Perspectives

This paper highlighted the problematic challenges faced by Phosphogypsum Tunisian authorities. The wet and dry embankments show two different mechanical behaviors. The previous studies make clear a better slope stability and fewer settlements and displacements for PG deposited via wet process where the embankment can be elevated to almost 70 m and can reach 100 m once reinforced by geotextile layers. In contrast, dry process results in a damaged embankment profile, important settlements and horizontal displacements, and cannot exceed 100 m in height without being reinforced by HDPE. Hence, wet process is well recommended to ensure better interaction between the embankment and the ground surface and to allow stocking almost 76% over of PG without the risk of getting damaged once important heights are reached.

Added to the industrial benefits of the wet process, this indicates that the TCG factory of the Skhira City would rather convert its deposition process into the wet one. This means that the existing embankment of the Skhira City is expected to receive at its top a new stack deposited using the wet process such as slurry.

Land optimization is essential and the deposition of PG slurry under the existing dry embankment is an eventual solution unless no new deposition site could be provided in very brief delays. It must therefore be recognized that the mechanical behavior of the dry embankment of the Skhira City should be characterized once it is a support for the new stack deposited using wet process.

Based on the previous studies, nothing guarantees that the dry embankment can really support this slurry without getting destroyed as until now there is no idea about neither the deep of the cracks covering its top nor if they are connected to each other. The drillings occurred in the study of 2007 show the presence of some water horizons in the middle of the dry embankment. Field persons from the TCG reported that cavities full of water were discovered in the embankment body and that the water content there was superior to the content at the moment of rejecting phosphogypsum which makes one wonders if the water infiltration in the existing embankment could make worse the situation.

Furthermore, when dealing with the wet process, there must be at least two decantation basins essential for the wet method functioning. In dry process, experience shows that a bulldozer cannot move above phosphogypsum just rejected from the filter, it has to wait for 2 or 3 days till the PG hardens and during this period it becomes yet resistant to be flattened. Even if, it is not evident that the deposited water will follow the gravity and escape from the drainage point, knowing that the support is dry and porous compared to a wet process deposited basis.

The simulation of the mechanical behavior of the complex dry embankment–slurry as support–deposit is required to characterize the problem and judge if any reinforcements are required before applying this solution to overcome this challenge unless no new deposition site could be provided. The maximum height that can reach the eventual resulting embankment remains open to question.

References

Ajam, L., Ouezdou, M.B., Felfoul, H.S., El Mensi, R.: Characterization of the Tunisian phosphogypsum and its valorization in clay bricks. Constr. Build. Mater. **23**(10), 3240–3247 (2009)

Belaiba, A., Felfoul, H.S., Bedday, A., Ouezdou, M.B.: Valorisation du phosphogypse dans les briques de construction. In: Proceedings of Colloque Matériaux, Sols et Structures MS2 2004 (2004)

Bouassida, M.: Étude de stabilité du terril de phosphogypse de Sfax, Rapport d'étude géotechnique, Université de Tunis El Manar, ENIT (2012)

Chaari, A.: Stabilité d'un terril de phosphogypse. Projet de fin d'études, ENIT (2013)

Choura, M., Keskes, M., Chaari, D., Ayadi, H.: Study of the mechanical strength and leaching behavior of phosphogypsum in a sulfur concrete matrix. IOSR J. Environ. Sci. Toxicol. Food Technol. **9**, 8–13 (2015)

Felfoul, H.S., Clastres, P., Carles, G.A., Ouezdou, M.B.: Amélioration des caractéristiques du phosphogypse en vue de son utilisation en technique routière. Waste Sci. Tech. **28**, 21 (2002)

Felfoul, H.S., Clastres, P., Ouezdou, M.B., Carles, A.G.: Caractéristiques et Propriétés du Phosphogypse des Terrils de Sfax (Tunisie): influence de la Durée de Stockage. In: Proceedings of Colloque IREX, 25–26 Novembre 2003, Auscultation, Diagnostic et Evaluation des Ouvrages (2003)

FNAC for environment: projet d'aménagement et d'exploitation d'une décharge humide de phosphogypse à Skhira: Etude d'impact sur l'environnement (2012)

Fuleihan, N.F.: Phosphogypsum disposal-the pros & cons of wet versus dry stacking. Proc. Eng. **46**, 195–205 (2012)

Gorlé, D. et Reichert.: Le Phosphogypse comme Matériau de Remblai Routier, CR 28/85, Centre de Recherches Routières, Bruxelles (1985)

Hilton, J.: Phosphogypsum (PG): uses and current handling practices worldwide. In: The 25th Annual Lakeland Regional Phosphate Conference, 13–14 October 2010, Chairman, Aleff Group, Lakeland FL, London (2010)

Living with danger: in investigation of phosphogypsum pollution in the phosphate fertilizer industry, Sichuan Province, China. http://www.greenpeace.org/eastasia/Global/eastasia/publications/reports/food-agriculture/2013/Living%20with%20Danger%20report.pdf

International Atomic Energy Agency IAEA—Vienna. Radiation protection and management of norm residues in the phosphate industry (2013). http://www-pub.iaea.org/MTCD/publications/PDF/Pub1582_web.pdf

International Maritime Organisation, Convention on the Prevention of Marine Pollution by Dumping of Wastes and Other Matter, 1972, as amended (2006)

Moalla, R., Gargouri, M., Khmiri, F., Kamoun, L., Zairi, M.: Phosphogypsum purification for plaster production: a process optimization using full factorial design. Environ. Eng. Res. **23** (1), 36–45 (2017)

Sahu, S.K., Ajmal, P.Y., Bhangare, R.C., Tiwari, M., Pandit, G.G.: Natural radioactivity assessment of a phosphate fertilizer plant area. J. Radiat. Res. Appl. Sci. **7**(1), 123–128 (2014)

TCG: Phosphogypsum management GCT experience evaluation. In: Proceedings of Phosphogypsum: Challenges and Opportunities Workshop, 23–25 April, Tunis, Tunisian, pp. 522–528 (2014)

URIG: Étude de stabilité du terril de phosphogypse de Skhira, Université de Tunis El Manar, ENIT (2007)

Valverde-Palacios, I., Valverde-Espinosa, I., Fuentes García, R., & Martín Morales, M.: Geotechnical risk and environmental impact: the stability of phosphor-gypsum embankments in SW Spain. Electron. J. Geotech. Eng. **16** (2011)

World Population Prospects: United Nations, Department of Economic and Social Affairs, File POP/2: Average annual rate of population change by major area, region and country, 1950–2100, July 2015

New Wales—Gypstack extension project description, Mosaic co phosphate fertilizer facility – Florida. WWW.Reuters.com

Zairi, M., Rouis, M.J.: Impacts environnementaux du stockage du phosphogypse à Sfax (Tunisie). Bulletin-laboratoires des ponts et chaussées, pp. 29–40 (1999)

Variation of Some Hydrogeochemistry with CBR of Residual Soils from North-Central Nigeria: Impact of Underlying Lithology

Tochukwu A. S. Ugwoke[1(✉)] and Salome H. D. Waziri[2]

[1] Michael Okpara University of Agriculture, Umuahia, Nigeria
tcugwoke@yahoo.com
[2] Department of Geology, Federal University of Technology, Minna, Nigeria

Abstract. This work was done in an area lying between latitudes 9°05¹N and 9°36¹N and longitudes 6°01¹E and 7°00¹E which is underlain by Pan-African Nigerian Basement Complex gneiss/schist and granite and Campanian Bida Sandstone. A total of 18 water and 18 soil samples were collected from water wells and soil pits respectively occurring in the 3 lithologic terrains. The water samples were subjected to atomic absorption spectrophotometer to determine the concentrations of the cations - Ca^{2+}, Na^{+} and Mg^{2+}. The water samples were also tested with digital titrator to ascertain the total hardness and with EC meter to determine the electrical conductivity. The soil samples were subjected to sieve analysis to determine the soil types and to California bearing ratio (CBR) test to determine the CBR of the soil. Results of the atomic absorption spectrophotometer reveals that concentration of Ca^{2+}, Na^{+} and Mg^{2+} in the water samples ranges from 5.2 to 45.3 mg/l, 1.0 to 22.5 mg/l and 0.2 to 23.9 mg/l respectively; results of the digital titrator reveals that the water total hardness ranges from 11.0 to 184.0 mg/l while results of the EC meter reveals that the water electrical conductivity ranges from 20.0 to 640.0μS/cm. The sieve analysis reveals that all the soil are coarse-grained while the CBR test reveals that wet CBR of soils ranges from 1.45 to 132.42%. Result of the tests show that water samples collected from gneiss/schist terrain have relatively high concentration of the cations - Ca^{2+}, Na^{+} and Mg^{2+}, total hardness and electrical conductivity while the soils collected from same (gneiss/schist) terrain have relatively low wet CBR. The reverse situations occurred in water and soil samples collected from sandstone terrain. Water and soil samples collected from granite terrain have relatively intermediate values. This work has shown that the inverse variation existing between the groundwater hydrogeochemistry and soil CBR is controlled by the mineralogy and texture of the underlying lithology that formed the soil and hosting the groundwater.

Keywords: California bearing ratio · Gneiss/schist · Granite · Sandstone Cation concentration · Terrain

S. Hemeda and M. Bouassida (Eds.): GeoMEast 2018, SUCI, pp. 105–113, 2019.
https://doi.org/10.1007/978-3-030-01941-9_8

1 Introduction

Hydrogeochemistry is primarily made up of the ionic concentration of groundwater. Ionic concentration is a manifest of the geochemistry and weatherability of the lithology (aquifer) bearing the groundwater. For instance, works by Hanshaw and Back (1979) and Datta and Tyagi (1996) had shown that that concentration of Ca^{2+} and Mg^{2+} in groundwater results, primarily, from the gradual weathering of carbonate minerals (like dolomite and calcite) and alkaline earth silicate minerals (like anorthite and pyroxene) making up the aquifer. The authors concluded that the geochemistry of groundwater is controlled by the mineralogy of the aquifer. Groundwater contained in aquifer of complex mineralogy and high weatherability is, therefore, expected to have higher concentration of ions than groundwater contained in aquifer of few mineralogy and low weatherability.

Most often, parts of this water-bearing lithology (aquifer) exposed to weathering agents degrade to residual soil. Groundwater contained in the aquifer can, however, affect the overlying soil through capillarity - a process which varies with the soil grain size and groundwater table (depth) at the time of the rise (Pal and Varande 1971; Garg 2011). The presence of some cations in capillary water has adverse effect on the soil mechanical stability. For example, the concentration of Fe^{2+}/Fe^{3+} in capillary water, emanating from groundwater, enhances the ferroginization and lateralization of the soil while ion exchange reaction of Ca^{2+}, Mg^{2+}, K^{+} and Na^{+} in soil water results to the concentration of Na^{+} in the soil and reduction in its stability (Hiscock 2005; Rosfjord et al. 2007; Lori 2007). Also, the suitability of the formed soil to satisfactorily serve for civil engineering purposes like pavement sub-grade, base/sub-base or filling materials has always being assessed based on some its (soil) geotechnical properties like California bearing ratio (CBR) and shear strength (FMWH 1997; ASTM D3080/D3080 M-11 2011; ASTM D1883-16, 2016;). CBR is a check of soil mechanical stability while shear strength is used to determine the resistance of soil to shearing stress.

As same water-bearing lithology (aquifer) can weather into residual soil, it is envisaged that there exists relationship(s) between the geochemistry of groundwater and the mechanical stability of the soil. This work evaluates such relationship using Ca^{2+}, Na^{+} and Mg^{2+} concentrations of the groundwater and wet California bearing ratio of the overlying residual soil. Findings of this work shall be useful in ascertaining the primary causes of variation in soil mechanical stability and the extent to which this mechanical stability varies with both the groundwater geochemistry and underlying lithology.

2 Regional Geology and Hydrogeology

The study area is underlain by gneiss, schist and granites of the Pan-African Nigerian Basement complex and Bida Sandstone of the Campanian Bida Basin. According to Rahaman (1988) and Obaje (2009), Pan-African Basement Complex formed as a result of global Pan-African tectonic event giving rise mostly to gneiss, schists and granites. Mineralogically, the gneiss is composed mostly of orthoclase, biotite, quartz, microcline and andesine/oligoclase; the schist is composed mostly of biotite, muscovite, quartz, hornblende and chlorite while the granite is composed mostly of quartz, muscovite, biotite and plagioclase (Obiora 2005; Okunlola et al 2009). Most often gneiss and schist, which are foliated rocks, co-occur. Thus, in the present work, the two rocks (gneiss and schist) shall be taken as one unit while granite shall be taken as a different unit. These Basement Complex rocks host groundwater within their fractured and weathered zones.

Bida Basin is a NW-SE extension Basin that formed during the Neocomian/early Gallic Epoch formation of the Benue Trough. According to Ofoegbu (1983), the Benue Trough experienced a Santonian tectonic event that resulted to fracturing, uplifting, folding and eventual formation of the Bida Basin. In Campanian Stage, Bida Basin received its first phase sediment – Bida Sandstone, which is composed mostly of poorly-sorted pebbly arkosic and quartzose friable sandstone (Nwachukwu 1972; Ojo 2012). Bida Sandstone is an unconfined aquifer of good porosity and primary permeability.

3 Sampling and Analyses

3.1 Sampling

Geological mapping of the study area augmented with information from literature revealed that the area is underlain by three lithologies namely: sandstone, granite and gneiss/schist as shown in Fig. 1. Brief petrogenesis and petrography of the lithologies have already been explained. Within the three lithologic terrains, a total of 18 water wells were located and water sample (coded w1-w18), were collected from each of the water wells. Collection of water sample was done early morning when there is maximum recharge, zero abstraction and minimum contamination of the groundwater due to abstraction. Each of the water samples was collected using a clean labelled 1 L plastic water bottle which was double-rinsed with the same water to be sampled before the collection to avoid contamination from the bottle. The samples were transported to the laboratory within 24 h for assessment of some major cationic concentration and physiochemical properties.

At close lateral range from each water-well, a soil pit was dug to depth range of 50 to and 70 cm from which soil sample was collected. The pits were dug to this depth range to ensure that the soil samples to be collected were in situ residual soil. Any pit containing up to 50% of grains greater than 60 mm (gravel) was abandoned and another pit dug in its replacement from which soil sample was collected. Thus, a total of 18 soil samples (coded x1 – x18) were collected and preserved in nylon bag for the sieve analyses and CBR tests.

3.2 Analyses

The cations determined in the water samples are calcium (Ca^{2+}), magnesium (Mg^{2+}), sodium (Na^{+}) while the physiochemical properties determined were total hardness and electrical conductivity. The concentration of the cations - Ca^{2+}, Mg^{2+} and Na^{+} were determined using Perkin – Elemer AAS 3110 and following the atomic absorption

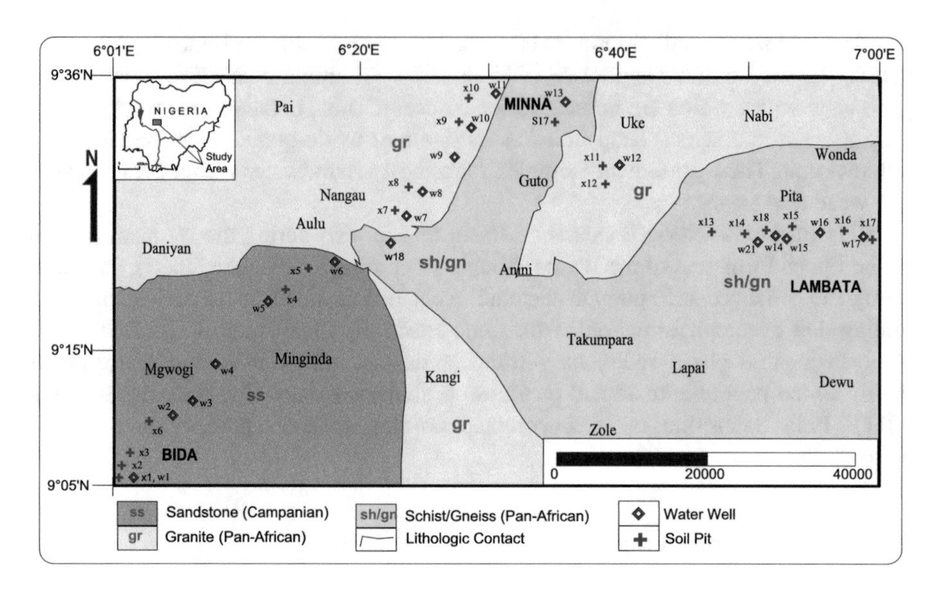

Fig. 1. Geological map of the study area showing locations of water wells and soil pits

spectrophotometery method described in SMEWW (1980) and Smith (1993). The total hardness (TH) was determined with Hach digital titrator while the electrical conductivity was measured using EC meter.

The soil samples were subjected to wet sieving following the BS 1377-2 (1990) standard and subsequently classified following the Unified Soil Classification System to ascertain the soil types. The soil samples were further subjected to California bearing ratio (CBR) test following ASTM D1883-16 (2016) standard to determine the wet CBR of the soils.

4 Results and Discussion

4.1 Water Ionic Concentration and Physiochemical Properties

Concentrations of Ca^{2+}, Mg^{2+} and Na^{+} in the water samples collected from the three lithologic terrains are shown as chart in Fig. 2 while their total hardness and electrical conductivity are shown in Fig. 3.

Figure 2 reveals that the cation concentrations vary according to the lithologic terrain from which the water samples were collected. Generally, the concentrations of the cations are highest in water collected from gneiss/schist terrain; intermediate in water collected from granite terrain and lowest in water collected from sandstone terrain. Following works by Garrels (1976); Hanshaw and Back (1979); Datta and Tyagi (1996) that geochemistry of groundwater is a reflection of the aquifer mineralogy and weatherability, the observed variation of the cations' concentrations amongst water samples from the three terrains can be explained in two ways. First is that, as was stated earlier, gneiss/schist and granite of the Nigerian Basement complex contain more

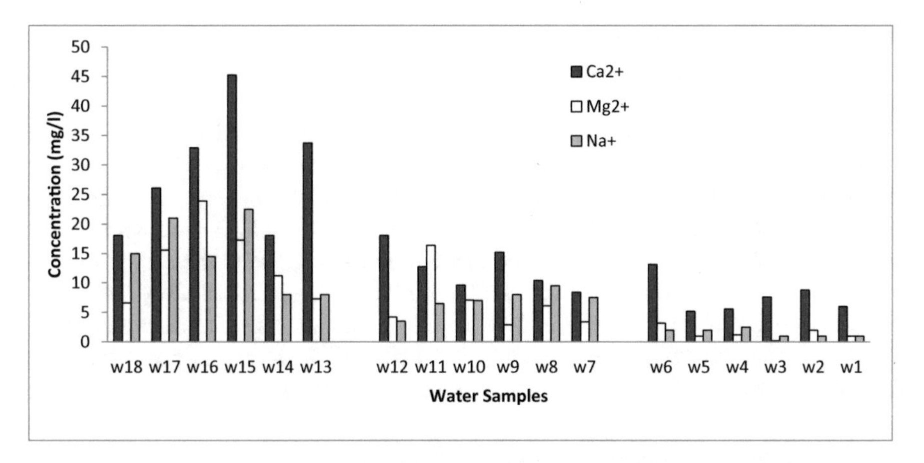

Fig. 2. Ca^{2+}, Na^{+} and Mg^{2+} concentrations in the analyzed water samples (w1-w6 is from sandstone terrain; w7-w12 is from granite terrain; w13-w18 is from gneiss/schist terrain)

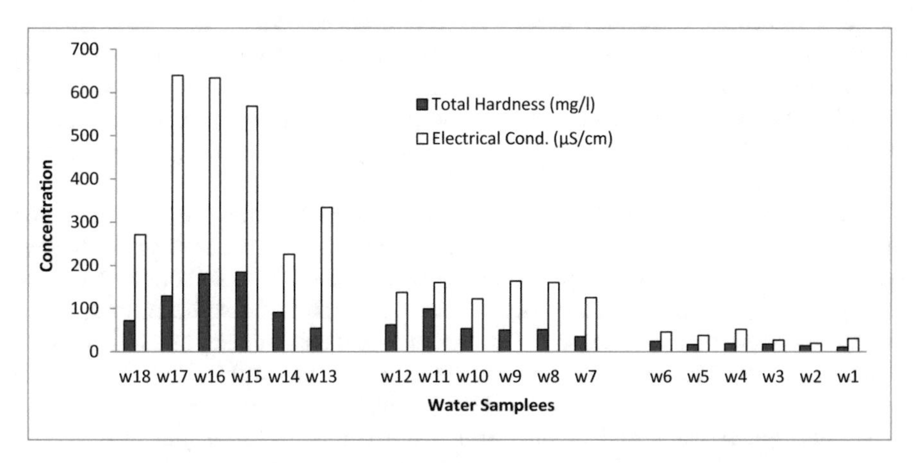

Fig. 3. Total Hardness and Electrical conductivity of the water samples (w1-w6 is from sandstone terrain; w7-w12 is from granite terrain; w13-w18 is from gneiss/schist terrain)

minerals than the Bida Sandstone. The high concentration of Ca^{2+}, Na^{+} and Mg^{2+} in samples collected from gneiss/schist and granite terrains relative to samples collected from Sandstone is attributed to the high concentration of low chemically stable plagioclase and hornblende in gneiss/schist. The friable Bida Sandstone dissociated significant amount of its Ca^{2+}, Na^{+} and/or Mg^{2+} rich minerals during its sedimentation process and so has little of these cations still left for dissociation into its groundwater. As this Sandstone had been described as arkosic, it may be richer in K^{+} which is chemically more stable than the other three cations. Secondly, the high susceptibility of the gneiss/schist aquifer to weathering (due to its foliated texture) and eventual dissociation of its mineral components to individual ions explains the higher concentration of cations in water collected from gneiss/schist terrain than water from granite terrain.

Hence, the mineral compositions and textural characteristics of the water-bearing lithology (aquifer) control the cationic concentration of the analyzed groundwater samples.

Figure 3 clearly shows that the total hardness and electrical conductivity of the water samples follow the same trend as their cationic concentration. It is factual that water hardness is controlled by the concentrations of Ca^{2+} and Mg^{2+} while work by Atkins and de Paula (2006) reveals that the electrical conductivity is controlled by both concentration and mobility of cations contained in the water. According to the author, mobility of cations in water follows the order: $\mu X^{+} > \mu Y^{+} > \mu Z^{+} > \mu Ca^{2+} > \mu Mg^{2+} > \mu Na^{+} > \mu^{W^{+}}$ where μ is mobility; X, Y, Z and W are cations not analyzed in the present work. Thus, the higher the concentration of Ca^{2+}, Na^{+} and Mg^{2+} in the groundwater, the higher the water hardness and electrical conductivity and vice versa.

4.2 Geotechnical Properties of the Soil Samples

The types and wet California Bearing Ratios (CBRs) of soils occurring in the three lithologic terrains are shown in Table 1 and Fig. 4 respectively. The table shows that the soils occurring in the three terrains are coarse-grained soils - sandy and gravelly soils. Soils occurring in the sandstone terrain are totally sandy soils while soils occurring in the gneiss/schist and granite are both sandy and gravelly soils. Understandably, this soil type variations is simply due to the underlying lithology; friable sandstone weather directly into sandy soils while gneiss/schist and granites weather into gravelly soils and subsequently to sandy soils.

Figure 4 shows that the CBR of the soil also vary according to the underlying lithology that formed the soil. Soils that formed from the sandstone have the highest CBR; soils that formed from granite have intermediate CBR while soils that formed from gneiss/schist have the least CBR. Again, this variation is attributed to the mineralogical composition and textural characteristics of the underlying lithologies which formed the soils. The Bida sandstone, which is composed mostly of relatively stable quartz (quartzose) and k-feldspars (arkosic), has higher mechanical and chemical stability than gneiss/schist and granite, which are composed of biotite, plagioclase, chlorite and hornblendes. Consequently, soils that formed from the Bida Sandstone have higher mechanical stability (CBRs) than the soils that formed from the gneiss/schist and granite. On the second hand, granite, which has graphic texture, is more stable than gneiss/schist, which has foliated texture. Therefore, gneiss/schist weathers to soil faster and the formed soil degrades having less mechanical stability (CBR) than granite.

4.3 Variation of the Soils' CBR with the Groundwater Cationic Concentrations

From our discussion so far, it can be seen that both the concentrations of the analyzed water cations – Ca^{2+}, Na^{+} and Mg^{2+} and the CBR of the soil vary according to the underlying lithology, which formed the soils and hosts the groundwater. Gneiss/schist lithology/aquifer resulted to water of relatively high Ca^{2+}, Na^{+} and Mg^{2+} and forms

Table 1. Classification of the analyzed soil samples

s/n	Terrain	Sample code	Unified Soil Classification System (USCS)	
			USCS Type	Description
1	Gneiss/schist	x18	GW-GM	Well-graded gravel with silt binder
2	Gneiss/schist	x17	SW	Well-graded sands, gravelly sands, little or no fines
3	Gneiss/schist	x16	SW-SM	Well-graded sand with silt binder
4	Gneiss/schist	x15	GW-GM	Well-graded gravel with silt binder
5	Gneiss/schist	x14	GP	Poorly-graded gravels, sandy gravels, little or no fines
6	Gneiss/schist	x13	GM	Silty gravels, gravel-sand-silt mixtures
7	Granite	x12	SW-SM	Well-graded sands with silt binder
8	Granite	x11	GW	Well-graded gravels, sandy gravels, little or no fines
9	Granite	x10	GW-GC	Well-graded gravel with clay binder
10	Granite	x9	GP-GM	Poorly-graded gravel with silt binder
11	Granite	x8	SW	Well-graded sands, gravelly sands, little or no fines
12	Granite	x7	GW	Well-graded gravels, sandy gravels, little or no fines
13	Sandstone	x6	SW	Well-graded sands, gravelly sands, little or no fines
14	Sandstone	x5	SP	Poorly-graded sands, gravelly sands, little or no fines
15	Sandstone	x4	SP-SC	Poorly-graded sands with clay binder
16	Sandstone	x3	SP-SM	Poorly-graded sands with silt binder
17	Sandstone	x2	SP-SM	Poorly-graded sands with silt binder
18	Sandstone	x1	SP-SM	Poorly-graded sands with silt binder

soils of low CBR; granite lithology/aquifer resulted to water of intermediate Ca^{2+}, Na^{+} and Mg^{2+} and formed soils of intermediate CBR while sandstone lithology/aquifer resulted to water of relatively low Ca^{2+}, Na^{+} and Mg^{2+} and formed soils of high CBR. Thus, via a vis gneiss/schist, granite and sandstone lithologies and provided the same water-bearing lithology forms the soil, the concentration of groundwater Ca^{2+}, Na^{+} and Mg^{2+} has inverse relationship with the wet CBR of the soil.

As the cationic concentrations of Ca^{2+}, Na^{+} and Mg^{2+} control the total hardness (TH) and electrical conductivity (EC) of the water, it is also conventional that these physiochemical properties (TH and EC) will vary with CBR as the cationic concentrations do. Terrains containing groundwater of relatively high TH and EC are characterised with soils of low CBR and vice versa.

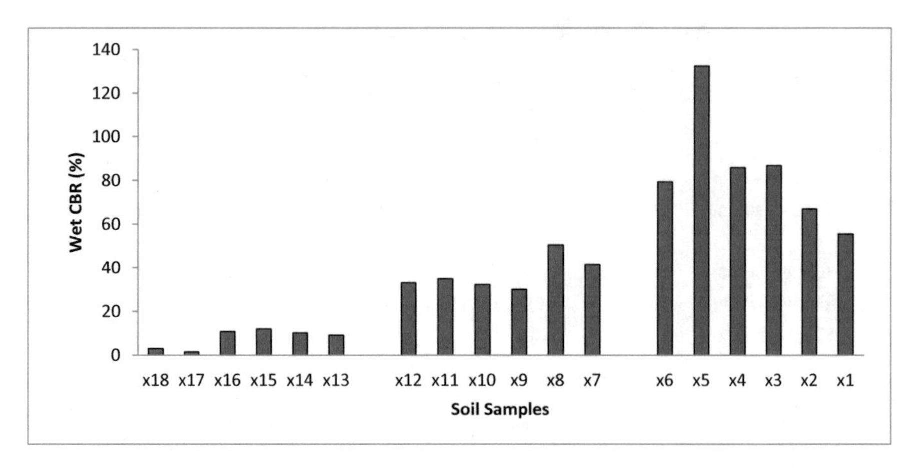

Fig. 4. Wet California Bearing Ratio (CBR) of the soil samples (x1-x6 is from sandstone terrain; w7-w12 is from granite terrain; w13-w18 is from gneiss/schist terrain)

5 Conclusions

The following conclusions are drawn from this work

1. Lithology hosting groundwater of relatively high Ca^{2+}, Na^{+} and Mg^{2+} concentrations formed soils of relatively low California bearing ratio (CBR) and vice versa.
2. Water and soil samples collected from gneiss/schist terrain have relatively high Ca^{2+}, Na^{+} and Mg^{2+} concentrations and low CBR respectively; water and soil samples collected from granite terrain have relatively intermediate Ca^{2+}, Na^{+} and Mg^{2+} concentrations and CRB; water and soil samples collected from sandstone terrain have relatively low Ca^{2+}, Na^{+} and Mg^{2+} concentrations and high CBR respectively.
3. The concentrations of Ca^{2+}, Na^{+} and Mg^{2+} in the groundwater and CBR of the soils are both controlled by the mineralogical composition and textural characteristics of the underlying lithology (aquifer).

References

ASTM D1883-16: Standard test method for california bearing ratio (cbr) of laboratory-compacted soils. ASTM International, West Conshohocken, PA (2016). https://doi.org/10.1520/d1883-16. http://www.astm.org/cgi-bin/resolver.cgi

ASTM D3080 / D3080 M-11: Standard test method for direct shear test of soils under consolidated drained conditions. ASTM International, West Conshohocken, PA (2011). https://doi.org/10.1520/d3080_d3080m-11. http://www.astm.org/cgi-bin/resolver

Atkins, P., de Paula, J.: Physical Chemistry. Oxford University Press (2006)

BS 1377-2: Methods of test for Soils for civil engineering purposes-Part 2: Classification tests (1990)

Datta, P.S., Tyagi, S.K.: Major ion chemistry of groundwater in Delhi area: chemical weathering processes and groundwater flow regime. J. Geol. Soc. India **47**, 179–188 (1996)

Federal Ministry of Works and Housing: General Specification for Roads and Bridges, vol. II, Federal Highway Department, FMWH: Lagos, Nigeria, pp. 317 (1997)

Garg, S.K.: Geotech Engnineering: Soil Mechanics and Foundation Engineering, 8th edn. Khanna Publishers, Daryaganj, New Delhi (2011)

Garrels, R.M.: A survey of low temperature water-mineral relations. Interpretation of Environmental isotope and hydrochemical data in groundwater hydrology, pp. 65–84. International atomic energy Agency, Vienna (1976)

Hanshaw, B.B., Back, W.: Major geochemical processes in the evolution of carbonate-aquifer systems. J. Hydrol. **43**, 287–473 (1979)

Hiscock, K.M.: Hydrogeology: principles and practice, p. 389. Blackwell Publishing Company (2005)

Lori, S.: Hazard identification for human and ecological effects of sodium chloride road salt. Final report prepared for State of New Hampshire Department of Environmental Services Water Division Watershed Management Bureau. I-93 Chloride TMDL Study (2007)

Nwachukwu, S.O.: The Tectonic evolution of the Southern portion of the Benue Trough Nigeria. Geol Mag. **109**, 411–419 (1972)

Obaje N.G.: Geology and Mineral Resources of Nigeria, Lecture notes in Earth Sciences 120, doi https://doi.org/10.1007/978-3-540-92685-91, C_ Springer-Verlag Berlin Heidelberg (2009)

Obiora, S.C.: Field descriptions of hard rocks with examples from nigerian basement complex, p. 44. Snaap Press Ltd, Enugu (2005)

Ofoegbu, C.O.: A model for the tectonic evolution of Benue Trough of Nigeria. Geol Rundschau. **73**, 1007–1018 (1983)

Ojo, O.J.: Depositional environments and petrographic characteristics of bida formation around share-pategi, Northern Bida Basin Nigeria. J. Geogr. Geology. **4**(1), 224–241 (2012)

Okunlola, O.A., Adeigbe, O.C., Oluwatoke, O.O.: Compositional and petrogenetic features of schistose rocks of Ibadan area, south-western Nigeria. Earth Sci. Res. J. **13**(2), 119–133 (2009)

Pal, D., Varade, S.B.: Measurement of contact angle of water in soils and sand. J. Indian Soc. Soil Sci. **19**(4), 339–343 (1971)

Rahaman, M.A.: Recent advances in the study of the basement complex of Nigeria. In: Geological Survey of Nigeria (ed) Precambrian Geol Nigeria, pp 11–43 (1988)

Rosfjord, C.H, et al.: Anthropogenically-driven changes in chloride indicate divergence in base cation trends in lakes recovering from acidic deposition. Environ. Sci. Technol. **41**(22), 7688–7693 (2007)

Standard methods for the Examination of Water and Wastewater (5th edn)-SMEWW: American Public Health Association–American Water Work Association–Water Pollution control Federation, Washington, D.C (1980)

G Shear Test to Determine Shear Characteristics of Coarse Grained Soils at Low Normal Stresses

V. Padmavathi(✉) and M. R. Madhav

Department of Civil Engineering, JNTUH College of Engineering, Hyderabad 500 085, India
vpadma70@gmail.com, madhavmr@gmail.com

Abstract. Shear strength is one of the most important properties of soil which controls various engineering aspects of geotechnical engineering such as bearing capacity, slope stability, pile capacity, etc. Therefore the determination of shear strength of soil is prerequisite for engineering purpose. The shear behaviour of soil is influenced by various factors such as drainage, compaction, confining stresses, etc. Hence, it is necessary to analyze the shear behaviour of soil that is appropriate to the field condition. Various tests such as direct shear, ring shear, tri-axial, unconfined compressive strength tests, etc., have been developed to determine the shear strength of soil. These tests are often operated at relatively high normal stresses. As soil behaviour at low normal stress is very different from that obtained using the above tests, a new test named as Gravity shear (G-Shear) test has been developed in which shear testing of soil can be performed at low normal stresses. Direct and G shear tests have been performed on sand at relative densities of 50% and 80% at different normal stresses. Results from these tests have been analyzed to contrast the shear behaviour of soil with the normal stress.

1 Introduction

Determination of shear parameters of soil is essential in Geotechnical Practice. The standard direct/box shear test has several limitations. The size of the sample in the standard shear box is very small, of the order of 60 mm × 60 mm because of which only sands with maximum particle size of about 5 mm can be tested. Large direct shear box of size 300 mm × 300 mm is available but suffers from a major problem of tilt of the top plate during test as the point of loading becoming eccentric with respect to the geometric centre of the box. The normal stress in routine testing ranges from 50 kPa to 200 kPa in the standard test while in the field the normal stresses can be much smaller at shallow depths in slopes particularly in shallow landslides. Mohr-Coulomb failure envelope of granular soils is non-linear with angle of shearing resistance being higher at low normal or confining pressures (Bolton 1986; Fannin et al. 2005). Gravity Shear Box (G Shear Box) has been developed in the laboratory to simulate possible field conditions that trigger the failure in shallow landslides. Senatore and Iagnemma (2011) studied the shear behavior of dry granular soil at low normal stress for application in

S. Hemeda and M. Bouassida (Eds.): GeoMEast 2018, SUCI, pp. 114–122, 2019.
https://doi.org/10.1007/978-3-030-01941-9_9

light weight robotic vehicle modeling. Lopes et al. (2014) have conducted inclined plane shear tests on geosynthetics with two types of granular soils of different size and grading.

2 Properties of Soils

Soil used in this study is sand. Grain size distribution, specific gravity, relative density, I_D, and direct shear tests on sand at 50 and 80% relative densities are conducted. The properties of soil are summarized in Table 1. The soil is poorly graded sand (SP).

Table 1. Properties of soil used

Properties of sand	
Gravel (%)	09
Sand (%)	90.5
Silt and clay (%)	0.5
Coefficient of curvature (Cc)	0.87
Uniformity coefficient (C_U)	2.18
Classification	SP
γ_{dmax} (kN/m^3)	17.6
γ_{dmin} (kN/m^3)	15.0
γ_d at I_D = 50% (kN/m^3)	16.3
γ_d at I_D = 80% (kN/m^3)	17.0
Angle of shearing resistance from direct shear test at I_D = 50% (°)	32
Angle of shearing resistance from direct shear test at I_D = 80% (°)	37.6

3 Methodology

Gravity Shear Box (G Shear Box) consists of a lower and an upper half of boxes which are subjected to relative transverse movement along a predetermined plane as during the direct shear test except that only gravity stresses act on the sample. The soil sample in the box is laterally confined and loaded by self weight only. During testing an upward force is applied near one of the bottom edges of box to facilitate tilt, $\alpha°$, there by the top half box moves over the bottom half box freely under the gravity. Normal force, W is due to weight of sand filled in the top half of box and nominal load applied over the sand. Area of cross section of the box is A. At failure the normal stress, $\sigma = W.\cos\alpha/A$ and shear/transverse stress, $\tau = W \sin\alpha/A$ can be calculated. Figure 1 shows the schematic diagram of stresses acting along the normal and transverse directions. The displacement of the sample in the transverse and the normal directions and the relative displacement of the box are measured by means of dial gauges.

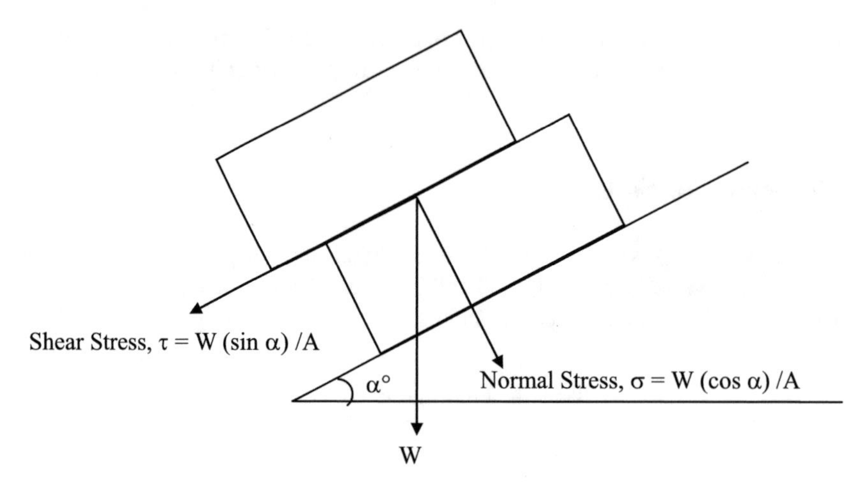

Fig. 1. Stresses acting along normal and transverse directions of gravity shear box

4 Experimental Setup

Gravity shear box test setup was fabricated with mild steel. It is square in shape and consists of two halves of 30 cm × 30 cm × 15 cm depth. The two halves of the box are held together by locking pins and placed on a steel frame. The lower half of the box is welded to a steel base plate of size 38 cm × 38 cm. The empty weight of the box along with the base plate is 25.41 kgs. One end of the steel plate abuts a vertical support facilitating lifting and rotation of the box assembly. A hydraulic jack of capacity 800 kN/m^2 is arranged at the bottom of the box to lift the box assembly at the other end and facilitates tilting of the box. The parts of the assembly unit are shown in Fig. 2. The two halves of the box are kept one over the other and temporarily fixed with bolts during sample preparation. The box is filled with soil (Sand). Steel plate is placed on the top of soil layer to place the dial gauge. Dial gauges of least count 0.01 mm and travel 100 mm are used for the measurement of transverse and normal displacements. One dial gauge is fixed to the lower half of the box through a magnetic stand to observe the transverse displacement of the top half of the box with respect to bottom one. Another dial gauge is mounted on to an L-shaped frame welded to the lower box and placed on the top of the soil to measure the normal displacement. The arrangement of dial gauges and L-shaped frame is shown in Fig. 3.

5 Procedure

The Gravity shear box is lifted up using a hydraulic jack. The normal and transverse displacements are measured at increasing angles of tilt. The top half of the box slides over the lower half of the box because of reduction of normal stress and increase in

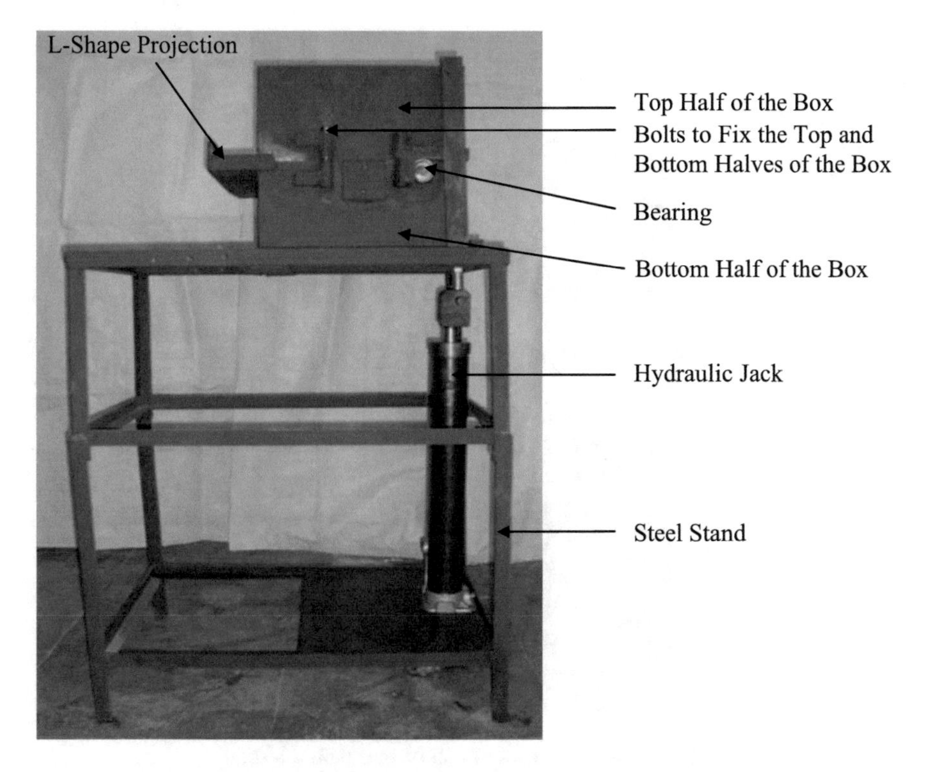

Fig. 2. Different parts of the gravity shear box

shear stress. Failure of the soil in the box (Figs. 4 and 5) along the interface between the halves of the box occurs at a particular angle of tilt is noted by rapid increase of transverse displacement. The inclination of the box, α, is calculated by measuring its movement in the vertical direction with the help of scale which is mounted on the frame. Friction between the edges of the box is minimized by polishing the surfaces and applying a lubricant.

6 Results and Validation

Poorly graded sand (SP) at different relative densities was used for testing. Figure 6 presents the normal and transverse stresses variation for 50% & 80% relative densities. This indicates the decrease in normal stress during the test and reaches to a minimum value at failure. The percentage decrease in normal stress at failure is 27% for both 50% & 80% relative densities, while it is constant in the direct shear test. In the same way, normal stress decreases marginally while transverse (or shear) stress increases significantly with tilt angle as the sample reaches the failure state in the G Shear Box test (Fig. 7). The variation of shear stress with normal stress at different tilt angle is plotted

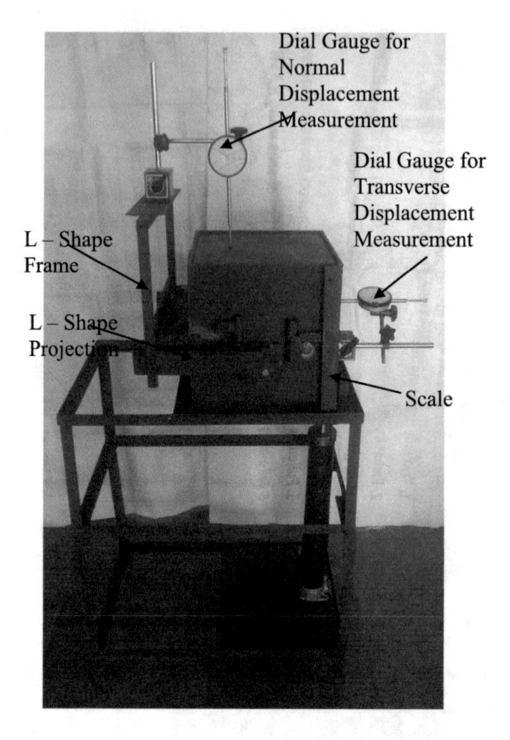

Fig. 3. Arrangement of dial gauges before testing

Fig. 4. Position of the gravity shear box just before failure

Fig. 5. Position of the gravity shear box before failure (top view)

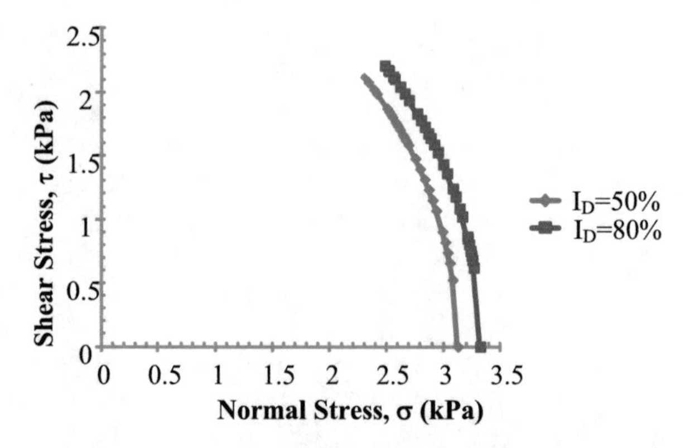

Fig. 6. Variation of normal and transverse stresses

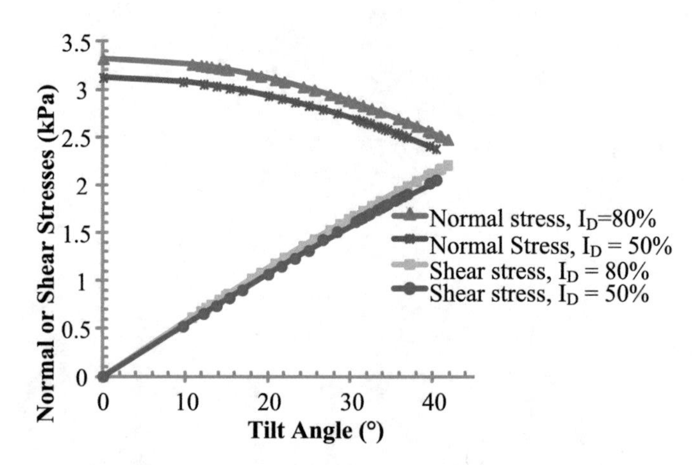

Fig. 7. Normal and shear stresses with respect to tilt angle

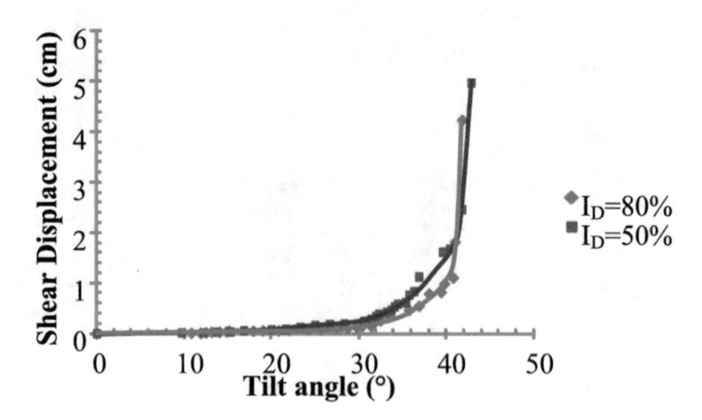

Fig. 8. Shear displacement vs tilt angle –effect of relative density

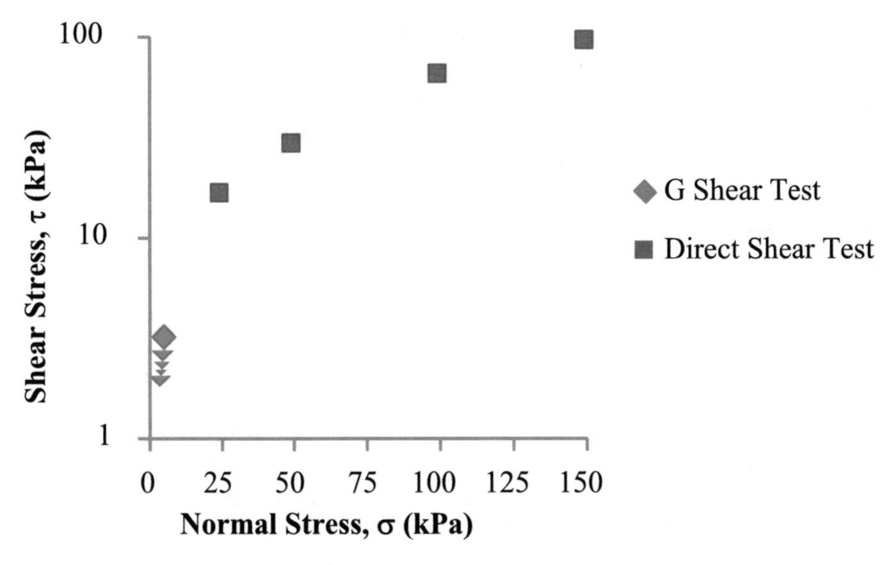

Fig. 9. Results of direct & G shear tests

Table 2. Comparison of results of direct & G shear tests

Relative density, I_D (%)	Angle of internal friction, ϕ (°)	
	Direct shear test	G shear test
50	32.0	41.0
80	37.6	42.0

for 50 and 80% relative densities as shown in Fig. 8. The failure point is identified at a certain value of tilt with sudden or rapid increase in transverse displacement. Failure is indicated by large shear displacement even with very small increase in tilt angle. The behaviour of soil at failure is catastrophic. The Mohr-Coulomb plot between normal and shear stresses depicted in Fig. 9 facilitates the estimation of angle of shear resistance of the soil. The results obtained with G Shear Box are compared with those from direct shear test (Table 2) and Fig. 9 which validates the non-linear nature of Mohr-Coulomb failure envelope at low confining pressures. The angle of shearing resistance of sand decreases with increase in normal stress as noted by Bolton (1986) and Fannin et al. (2005).

7 Summary and Conclusions

The soil in the proposed test is sheared by gravity only and under very low normal stresses. The angle of shearing resistance of the soil at low normal stress can be determined from proposed test. The Mohr-Coloumb failure envelope could be extended close to low normal stresses with the results supplemented from G-shear Test.

The normal stress decreases by about 27% for both 50% and 80% relative densities in G shear test whereas it is constant during direct shear testing. The angle of shearing resistance at low normal stresses in G shear test is 41° and 42° for relative densities of 50% and 80% respectively whereas it is 32° & 37.6° respectively for the same relatively densities at relatively high normal stresses in direct shear test.

Acknowledgment. J. Shivanandh, Postgraduate Student, JNTUH College of Engineering, Hyderabad helped in the fabrication of the equipment and testing.

References

Bolton, M.D.: The strength and dilatancy of sands. Geotechnique **36**(1), 65–78 (1986)

Fannin, R.J., Eliadorani, A., Wilkinson, J.M.T.: Shear strength of cohesionless soils at low stress. Geotechnique **55**(6), 467–478 (2005)

Lopes, M.L., Ferreira, F., Carneiro, J.R., Vieira, C.S.: Soil geosynthetic inclined plane shear behavior, influence of soil moisture content and geosynthetic type. Int. J. Geotech. Eng. **8**(3), 335–342 (2014)

Senatore, C., Iagnemma, K.D.: Direct shear behaviour of dry, granular soils for low normal stress with application to lightweight robotic vehicle modelling. In: Proceedings of the 17th ISTVS International Conference, Blacksburg, VA (2011)

Use of the Method of Concrete Lozenges to Strengthening the Slopes Stability: Assessment of the Safety Factor by the Finite Element Method

Latifa El Bouanani[1(✉)], Khadija Baba[1], and Latifa Ouadif[2]

[1] GCE Laboratory, High School of Technology-Sale, Mohammed V University in Rabat-Morocco, Rabat, Morocco
elbouanani.latifa@gmail.com, khadija_baba@hotmail.com
[2] GIE Laboratory, Mohammadia Engineering School, Mohammed V University in Rabat-Morocco, Rabat, Morocco
ouadif@gmail.com

Abstract. The major problem of unsaturated slope is the variation of their volumetric water content which is an important factor in the variation of mechanical parameters of soils into the subsurface. It can also influence the soil shear strength; consequently several methods were proposed to improve the unsaturated slope stability including the concrete lozenges technique. It is a new technique used to protect slopes. It constituted a non-continuous mesh mask on the slop, having the effect of collecting and transporting runoff water on the slope. Using the finite element method, the objective of this paper was to study both how the value of soil volumetric water content and the safety factor changed by the use of this new technique. Indeed, the safety factor is <1 for the shallow soil and increases with depth. Simulation and calculation results demonstrated that the proposed technique is capable of making the amount of soil masse affected, less than 1 m.

Keywords: Slope stability · Unsaturated slope
Finite element method · Concrete lozenges technique

1 Introduction

In discussing the slope stability, it is important to define what parameters control landsliding. In fact, slopes are affected by preparatory factors which can make them conditionally unstable such as climatic events. Indeed, we can define the safety factor of the slope as the ratio of the maximum load or stress that a soil can sustain to the actual load or stress that is applied.

Another phenomenon which is a result of the climatic events, especially intense rainfalls, is the water erosion. Water erosion is a natural phenomenon, which is defined as the total process of detachment, transport and deposition of solid particles from the soil surface due to rainfall, runoff or both. The risk of water erosion occurs in both drainage basins and road and rail slopes. It can also occur when the runoff capacity is

S. Hemeda and M. Bouassida (Eds.): GeoMEast 2018, SUCI, pp. 123–132, 2019.
https://doi.org/10.1007/978-3-030-01941-9_10

greater than the infiltration capacity due to soil saturation or to the formation of a bedding layer, named soil crusting, which minimizes the infiltration capacity of the soil (Elbouanani 2018).

In Hydrogeology, the unsaturated zone is the part of the subsoil located between the subsurface and the water table, which is an imaginary surface where the pore-water pressure is zero (Lam et al. (1987). In this zone, the pores of the soil are partially filled with water and air, unlike the saturated zone in which the entire porous system is filled with water (Casulli and Zanolli 2010). The flow of water into the soil can be described by the generalized Darcy law. Soil hydraulic conductivity, infiltration rate, water holding capacity and water table depth are soil hydraulic parameters which were required to study the flow of water through the slope. Soil hydraulic properties can be measured either directly in the field or in the laboratory (Alavi 2001). However, measurement of some soil hydraulic properties is costly.

Several methods were proposed to stabilize unsaturated slopes like concrete lozenges technique. Concrete lozenges are a preventive technique which proposes the creation of well-developed artificial gullies, notably concrete lozenges to frame the surface of exposed soil and to direct the runoff paths towards these gullies (Fig. 1) The advance of computer technology and the availability of the Finite Element Methods Geotechnical software provide engineers with sophisticated tools for analyzing Geotechnical problems on hand (Tjie-Liong 2014). The finite element formulation was also implemented into FEM Geotechnical software named Plaxis.

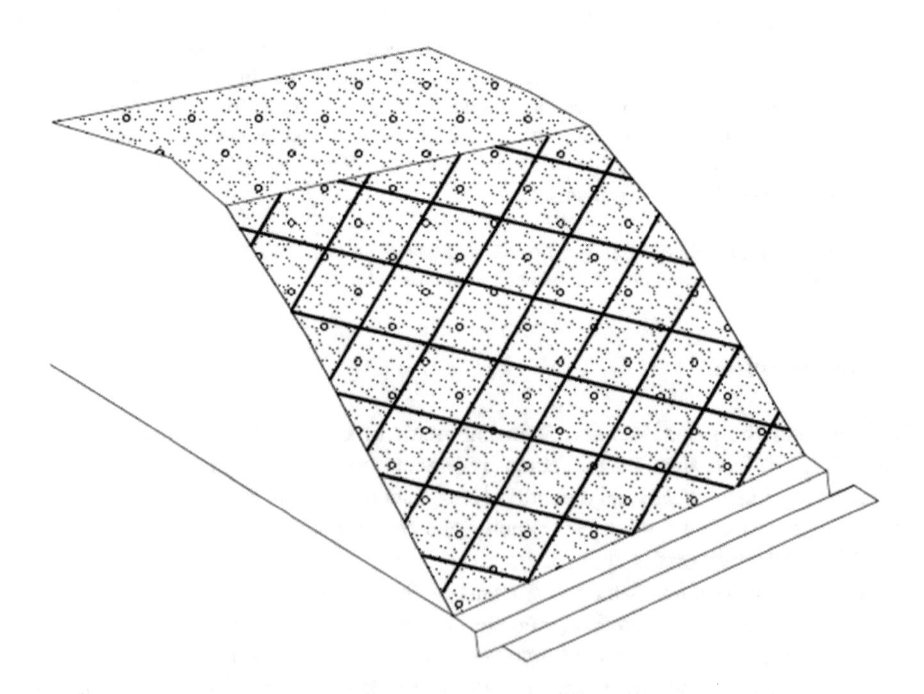

Fig. 1. Application of concrete lozenges' technique

Indeed, Slopes which surrounding the Taourirte-Nadour railway line, more precisely PK 94, due to their location in the Oriental Rif of Morocco, are subject to intense erosion, which necessitates periodic maintenance, generating significant costs. In this paper, Plaxis and Plaxflow FEM code calculation are used of the modeling of the slope geometry and the boundary conditions.

2 Theoretical Framework

Water has the tendency to flow through pores under the influence of a hydraulic gradient, which is defined by Bernoulli's theorem as:

$$h = \frac{v^2}{2g} + \frac{u_w}{\varrho_w g} + z \tag{1}$$

Where: h = total head, u_w = pore water pressure (KPa), z = elevation head above an arbitrary datum (m), ϱ_w = density of water (KN/m3), v = seepage velocity head (m/s) and g = acceleration due to gravity (m/s2).

In the unsaturated soil the seepage velocities are small so the velocity head can be neglected, and we can identify Bernoulli's theorem as:

$$h = \frac{u_w}{\varrho_w g} + z \tag{2}$$

The flow of water through a porous medium can be described using Darcy's law:

$$q = ki \tag{3}$$

Where q = discharge per unit area, i = potential gradient, and k = hydraulic conductivity (permeability).

Darcy's law is applied also for water flow in an unsaturated soil. Unsaturated soil is a tri-phasic medium so two phases, water and air, coexist in the pore channels of the soil. To describe the flow of water through a two-dimensional element of unsaturated soil, we need two partial differential equations:

$$\Delta q = \frac{\partial \theta_w}{\partial t} = -\frac{\partial}{\partial x}\left(k_x \frac{\partial h}{\partial x}\right) - \frac{\partial}{\partial y}\left(k_y \frac{\partial h}{\partial y}\right) \tag{4}$$

The hydraulic conductivity of the soil at all points in both the saturated and unsaturated zones is important to describe the flow in them. Van Genuchten (1980) and several describe an unsaturated flow model where the amount of water retained in the unsaturated zone of a particular soil depends on the soil water potential ψ (matric suction).

The water retention curve is the relationship between the water content θ, and the soil water potential ψ (matric suction).

The saturated coefficient of permeability can be obtained using simple field tests, while the unsaturated permeability is often expensive, difficult to conduct and the time-consuming. However, several researches have used models for the calculation of unsaturated permeability of a soil based upon soil water retention curve (SWRC) (Van Genuchten 1980).

Indeed, Van Genuchten (1980) has been described an equation for predicting the hydraulic conductivity of unsaturated soils. It's based on the Mualem's (1976a) model for predicting the hydraulic conductivity from knowledge of the soil water retention curve and the conductivity at saturation. Indeed, Irmay Irmay (1954) describe the relative liquid or gas permeabilities as a function of the degree of liquid saturation.

The effective degree of saturation can be described by the follow equation:

$$S_e = \frac{S - S_r}{S_s - S_r} \tag{5}$$

Where: s and r indicate respectively the saturated and the residual values of degree of saturation and the saturated permeability of the soil.

According to Van Genuchten (1980) the degree of saturation can also be expressed on terms of the pore water pressure of the soil as:

$$S(p_w) = S_{res} + (1 - S_{res})\left(1 + \left(\frac{\alpha p_w}{\rho_w g}\right)^n\right)^{-m} \tag{6}$$

$$m = 1 - 1/n \,\&\, 0 < m < 1$$

Where: α, n and m are the independent parameters of the soil.

Where: Sres indicate the residual value of the degree of saturation of the soil and pw is the water pressure.

The soil's void is filled both by water and air. The residual degree of saturation describes the amount of water remains permanently in the pores.

The relative hydraulic conductivity can be expressed on terms of the degree of saturation as:

$$K_{rel}(S) = S_e^{gl}\left[1 - \left(1 - S_e^{\frac{1}{m}}\right)^m\right]^2 \tag{7}$$

3 Study Case

The study presented in this paper consists in the application of the concrete lozenges technique for an unsaturated slope located in the PK94 of Taourirt-Nador railway line in Morocco. The unsaturated soil of the slope is simulated by an experimental and a numerical model to follow the influence of the effective permeability variation in depth on the slope stability, using the Plaxis and Plaxflow code calculation (FEM geotechnical software).

3.1 Modeling of Un Unsaturated Slope Stabilized by Concrete Lozenges Technique

The slope comprises 02 layers whose cumulative thickness is 8 m. The characteristics of each layer are given by the Table 1.

Table 1. Characteristics of the soil

Layer	$\Psi°$	E_{ref}	v
Marl	0	1800	0.33
Yellowish marl	0	2100	0.3

The modeling of the slope on the software takes into account the influence of the introduction of concrete lozenges on the variation of the degree of saturation of the soil. Indeed, as will be explained in the next part, the degree of saturation of the soil varies in depth. This variation has an influence on the hydraulic and mechanical parameters. At the software level, the part of the slope formed by marl is divided into sub-layers of

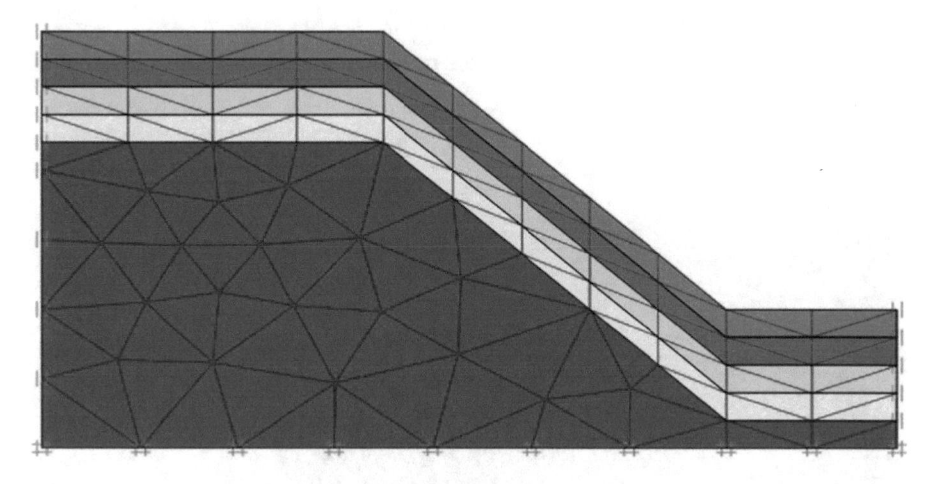

Fig. 2. Kinematic boundary conditions

1 m thick (Fig. 2), the difference between them is at the level of said parameters, namely:

- General properties : Υ_{sat} & Υ_{unsat}
- Permeability : K_x & K_y
- Strenght : C & φ

Table 2 has shown the values taken for each layer.

Table 2. The mechanical and hydraulic parameters of the layers

		Layer n°1	Layer n°2	Layer n°3	Layer n°4
General properties	$\Upsilon_{sat\ KN/m3}$	18	18	18	18
	$\Upsilon_{unsat\ KN/m3}$	18	17	16,5	16
Permeability	$K_{x\ m/day}$	0,15	0,147	0,143	0,14
	$K_{y\ m/day}$	0,15	0,147	0,143	0,14
Strength	$C_{\ KN/m2}$	5	10	15	17
	φ	25	27	30	30

NB: layer 1, 2 …. From top to bottom

As will be noticed the mechanical and hydraulic parameters of the layers is in relation with the degree of saturation of the soil.

The boundary conditions are as follows:

- The horizontal displacements along the X axis, and vertical along the Y axis on the bottom of the massif are all zero.
- The boundary conditions for the flow are deduced, by default, from the general level of the water table by the PLAXIS program. The PLAXFLOW program is used for unsaturated and transient flows.

Figures 2 and 3 respectively represent the mesh and the boundary conditions of the numerical model.

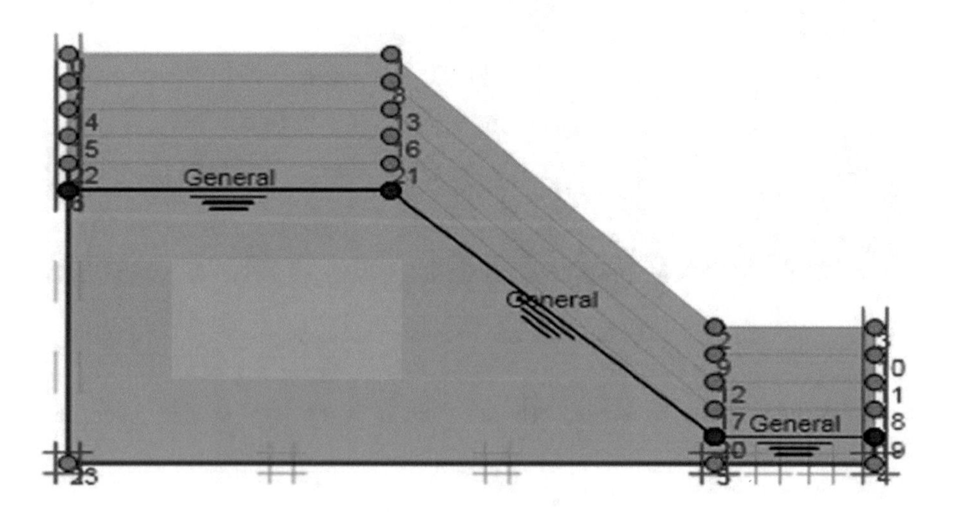

Fig. 3. Hydraulic boundary conditions

4 Results and Discussions

In this part, we study both the effect of introducing concrete channels on the slope and her effect on the water infiltration. As a result, the slope instability is assumed to be due to water infiltration from the soil surface and the instability zone is considered to be in a higher level and sufficiently far from the water table.

A reduced model and a dripper rainfall simulator were used to study the influence of the concrete lozenges technique on infiltration and runoff from bare, crusted dryland soils.

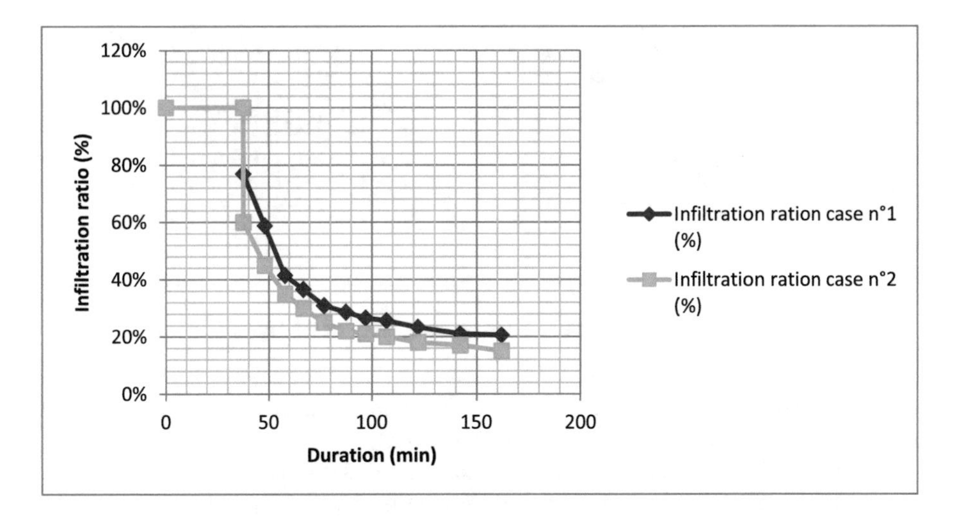

Fig. 4. Infiltration ratio as a function of time

Rainfall was applied at mean rain rates of 30 mm/h. For comparative purposes, a reduced model without protection was studied.

Infiltration as a Function of Time

The graph (Fig. 4) has shown that for the two cases; reduced model of slope with protection (case n°2) and without protection (case n°1), infiltration begun after a few times, when runoff begun after 40 min and the infiltration decreases over time.

It may be justified by:

The formation of a bedding layer, named soil crusting, which minimizes the infiltration capacity of the soil;

It can also occur when the runoff capacity is greater than the infiltration capacity due to soil saturation;

Water flows on the slope and collected through the concrete channels. Concrete lozenges channels have the ability to collect runoff water and transport it to the foot of the slope. It reduces the amount of water infiltrated. So it has the ability to preserving suction during rainfall in depths.

Degree of Saturation as a Function of Depth

Since the degree of saturation depends on the quantities of water infiltrated at depth, concrete lozenges allow the collection of runoff water and transport it before infiltration

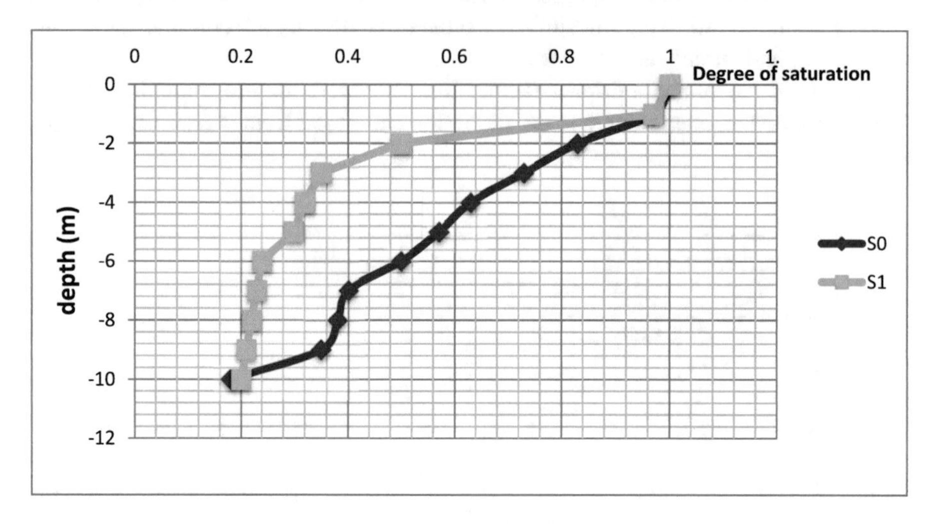

Fig. 5. Degree of saturation as a function of depth

into the ground, which justifies the difference between the degree of saturation at depth between the two cases; with and without protection.

The graph (Fig. 5) has shown that when stabilizing the slope by concrete lozenges, the degree of saturation is dropped to 100% for the shallow depths (1 m). After that infiltration has little influence on the soil saturation.

So, the effect of the concrete lozenges technique is significant in depths. In the shallow depth the matric suction is dropped to zero because of the saturation of the soil.

Where: S0 is the degree of saturation for a reduced model of a slope without protection and S1 is the degree of saturation for a reduced model of a slope stabilized by concrete lozenges channels.

Taking into account the values of the degrees of saturation. The modeling on Plaxis made it possible to properly assess the results given by the Fig. 6.

Fs as a Function of Depth

The calculation results for the three cases of the Security Factor indicated that the Security Factor is less than 1 for 0 to −1 m for the case n°1 and for 0 to −2,5 m for case n°2. After that Security factor increases with depth. So, concrete lozenges technique can make the amount of soil masse affected, less than 1 m.

The saturation of the soil in depth is a function of time and the rain's intensity.

For the first case (Case n°1) which represents the safety factor for a slope which is subject to rain and it is protected by concrete lozenges channels, the non-stability of the part of the slope lying between 0 and 1 m is linked to the saturation of the shallow soil before the start of the runoff.

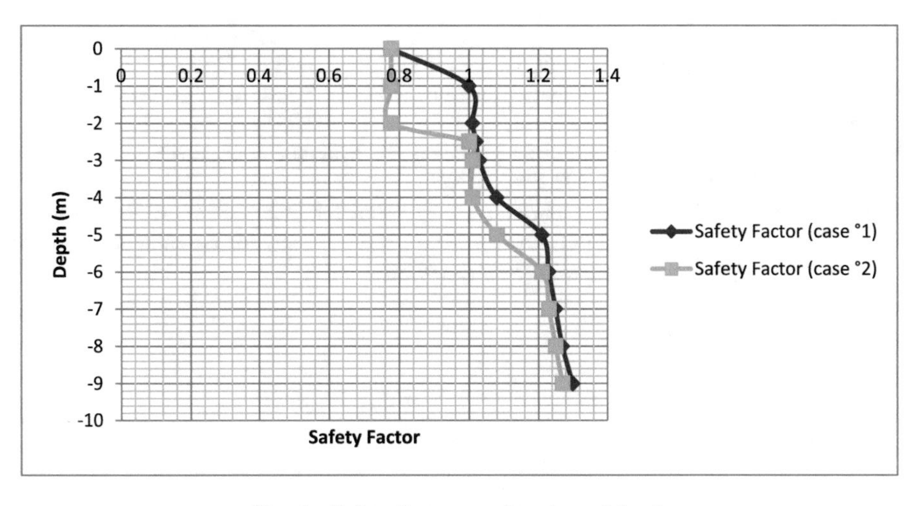

Fig. 6. Safety factor as a function of depth

For the second case (Case n°2) which represents the safety factor for a slope which is subject to rain and it is not protected. Water infiltration continues in deep because of the lack of a device for collecting water and because of the erosion of the shallow soil layer.

5 Conclusion

A numerical model of the slope is defined using Plaxis code calculations, to consider both the mechanical and the hydraulic effects of concrete lozenges channels on the slope stability.

Parametric study shows that after a simulated rainfall, concrete channels can stabilize the shallow soil of up to 9 m for a slope of 10 m of depth.

Since the hydrological model consider a 1d water flow, a direct comparison indicates that the concrete lozenges technique has a significant effect on the Security Factor, because of the induced variation on the soil hydraulic properties, especially the degree of saturation and the matric suction of the soil.

References

Lam, L., Fredlunda, D.G., Barbou, N.D.S.L.: Transient seepage model for saturated-unsaturated soil systems: a geotechnical engineering approach. Can. Geotech. J. **24**, 565–580 (1987)

Elbouanani, L., Baba, K. et al.: Concrete lozenges impact on the slope erodibility. In: MATEC Web Conference, 2nd International Congress on Materials & Structural Stability (CMSS-2017), vol.149, p. 02073 (2018). https://doi.org/10.1051/matecconf/201814902073

Alavi, G.: Estimation of soil hydraulic parameters to simulate water flux in volcanic soils. N. Z. J. For. Sci. **31**(1), 51–65 (2001)

Tjie-Liong, G.: Common mistakes on the application of plaxis 2D in analyzing excavation problems. Int. J. Appl. Eng. Res. **9**(21), 8291–8311 ISSN 0973–4562 (2014)

Irmay, S.: On the hydraulic conductivity of unsaturated soils. Trans. Am. Geophys. Un. **35**(3), 463–467 (1954)

Van Genuchten, M.T.: A closed form for predicting the hydraulic conductivity of unsaturated soils. Soils Sci. Am. Soc. **44**, 892–898 (1980)

Casulli, V., Zanolli, P.: A nested newton-type algorithm for finite volume methods solving richards' equation in mixed form SIAM J. Sci. Comput. Society for Industrial and Applied Mathematics, vol. 32, No. 4, pp. 2255–2273 (2010)

Modeling Landslides by the Finite Element Method: Application to an Embankment on a Railway in the Moroccan Rif

Ghizlane Ardouz[1(✉)], Khadija Baba[1], and Latifa Ouadif[2]

[1] LGCE Laboratory, High School of Technology-Salé, Mohammed V University in Rabat-Morocco, Rabat, Morocco
ghizlane.ardouz@gmail.com

[2] 3GIE Laboratory, Mohammadia Engineering School, Mohammed V University in Rabat-Morocco, Rabat, Morocco
ouadif@gmail.com

Abstract. Slope instability is one of the problems that often affects the stability of civil engineering infrastructure. These unstable slopes have often led to excessive damage resulting in high budgets to maintain these infrastructures.

The study area located in the Fez/Taza railway line is often subject to sliding phenomena leading to slope instabilities that require rapid and costly interventions. In fact, major landslide disturbances have occurred in the area's embankments. The observations on site show the following disorders: settlement at the platform, scouring, overturning of the guardrails ballast, beads, silting of gutters, ruptures of the descents of water. Becoming increasingly important these disorders may compromise the exploitation of these works or affect their safety. On each side of the embankment object of study, a geotechnical companion consisting of the realization of soundings with samples taken intact, and the establishment of various devices (piezometric test, inclinometer test, Core drilling coupled with pressuremeter tests). They revealed the existence of loose marls up to 12 m, followed by a more consistent yellow marl from 12 m to 22 m, and from 22 m to 28 m the marl becomes greyish and consistent. Inclinometers showed the existence of significant horizontal displacements in the direction of the slope at a dimension of −6 m compared to the coast of Natural Land.

The present study made it possible to compare on a real geometrical model the results of calculation of the displacement maximal by different methods: inclinimeters installed on site and methods of the finite elements. The results obtained with both methods are similar.

1 Introduction

Geomechanical models are practical tools for assessing potentially unstable slope stability, [1]. Their development is necessary in order to mitigate the destructive effects of the movements of ground. These numerical modeling tools are increasingly becoming an essential part of geotechnical expertise and can in particular help a better understanding of the physical processes and behavior of soils in a given problem [2]. The challenges of geomechanical analysis of landslides come from the specific

S. Hemeda and M. Bouassida (Eds.): GeoMEast 2018, SUCI, pp. 133–141, 2019.
https://doi.org/10.1007/978-3-030-01941-9_11

expertise required in modeling, but even more so because geo-mechanical modeling is part of an interdisciplinary process involving several stages involving both geological, hydrological and geomechanical models [2] (Fig. 1).

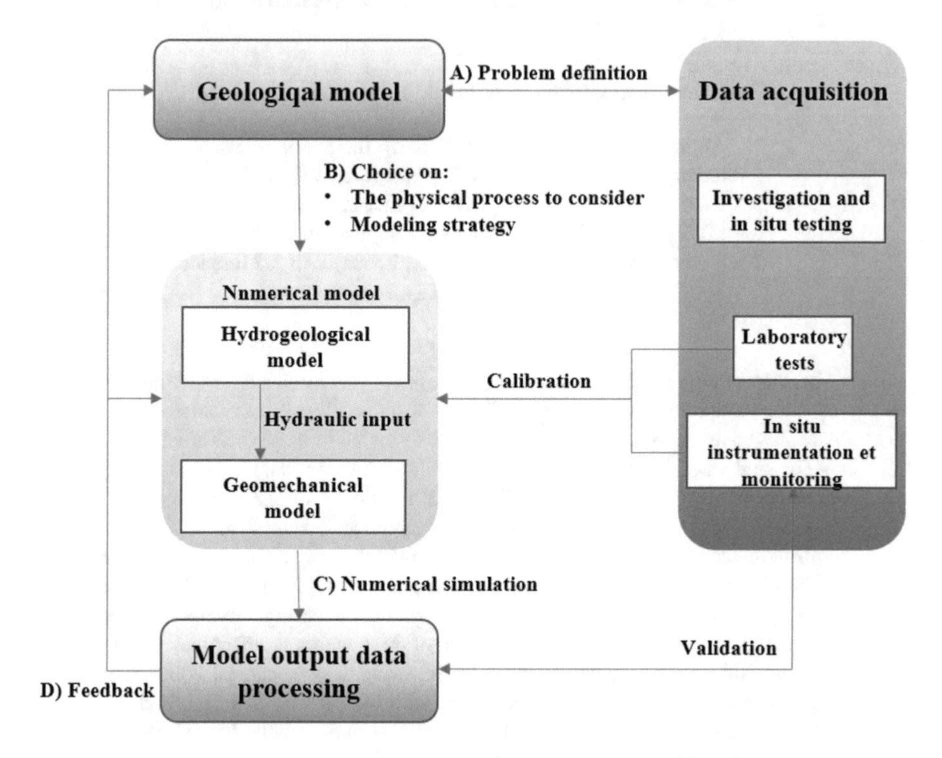

Fig. 1. Iterative and interdisciplinary process of landslide modeling [2].

2 Physical Processes in Slopes Subject to Rainfall

Generally, in natural slopes, there are two types of underground flow regimes. The first is a deep flow, mostly parallel to the surface of the slope. The second is superficial with positive water pressures or capillary suctions that depend on rainwater that infiltrates mainly vertically through the surface [3]. The nature of the movements, sometimes slow and rigid, sometimes fast and rather fluid, depends on a multitude of factors including the intensity of the precipitation and their duration, the capacity of dissipation of the pressures of water of the soil, the heterogeneity of the layers of soil at the site, the inclination of the slope, the type of soil, its grain size and its density and state of stress before the main rain event [2]. A flow parallel to the slope surface may settle after a rain event when the saturation front reaches a relatively impermeable substratum. In this case, a considerable volume of water is transported to the foot of the slip [2]. Several hydromechanical processes can have a destabilizing influence on a slope. The degree of saturation of the upper soil increases, reducing the capillary tension between the soil particles, weakening the soil that constitutes a slope. If the intensity of precipitation is greater than the capacity of the soil to dissipate pore water, surface runoff may occur, which may erode the slope [2].

3 Influence of Capillarity on Soil Resistance

Capillarity is a phenomenon that results from the surface tension of fluids. This tension develops at the interface of different materials. In soils, the capillary menisci retain the particles bound together, the phenomenon is called apparent cohesion (Fig. 2). The capillarity thus contributes to increasing the contact forces and improves the frictional resistance between the particles.

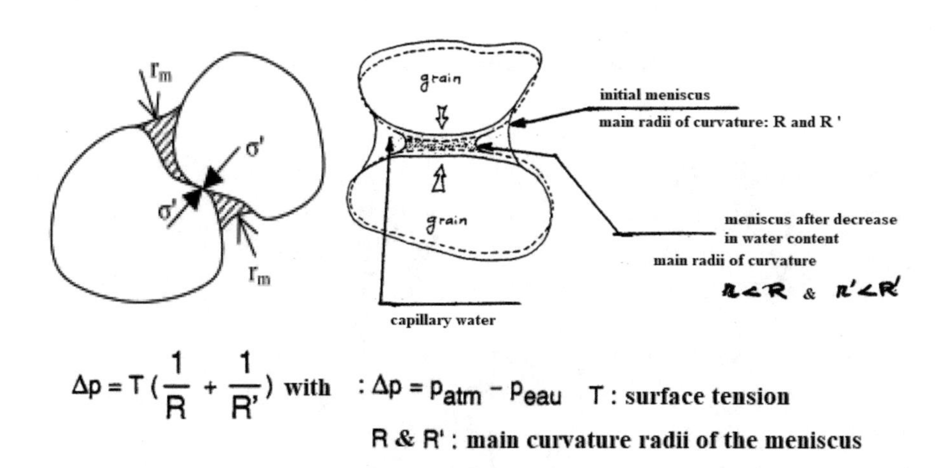

Fig. 2. Cohesion apparent (left) - and the principle of capillarity (right).

Suppose the soil is subjected to desiccation; the capillary water will evaporate into the atmosphere, the rays of the capillary menisci will decrease and we see from the Laplace formula (Δp) that it will result in greater capillary cohesion, and new grains came into contact, the adsorption cohesion will also increase (Fig. 3).

The mechanical strength of the clay must therefore increase, this is what can be observed on the curve which shows the variation of the simple compressive strength of a fine soil as a function of the water content.

4 Numerical Modeling of Landslide Due to Precipitation

On the railway line connecting Fez and Oujda, and 8 km from the city of Fez, a landslide occurred on a 20 m high embankment following heavy rainfall experienced by the region during the years 2009 to 2013 (Fig. 4). The geometric data of the site as well as the soil characteristics were used to create the finite element model shown in Fig. 5.

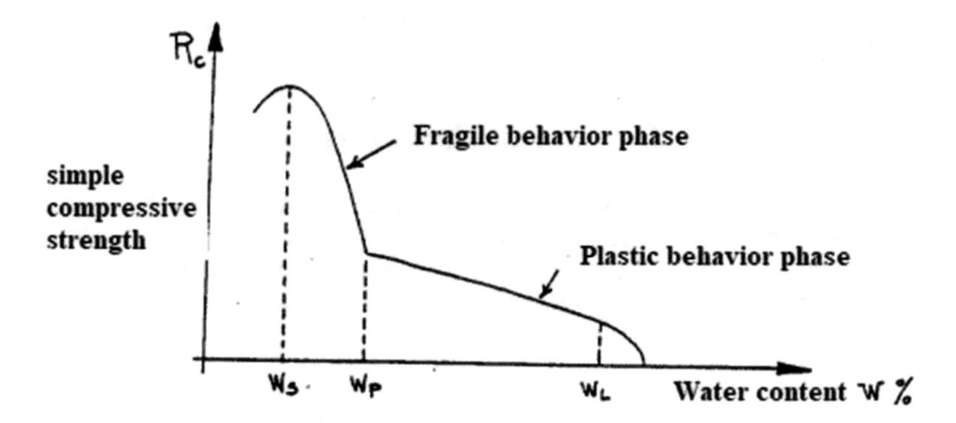

Fig. 3. Variation of the cohesion as a function of the water content.

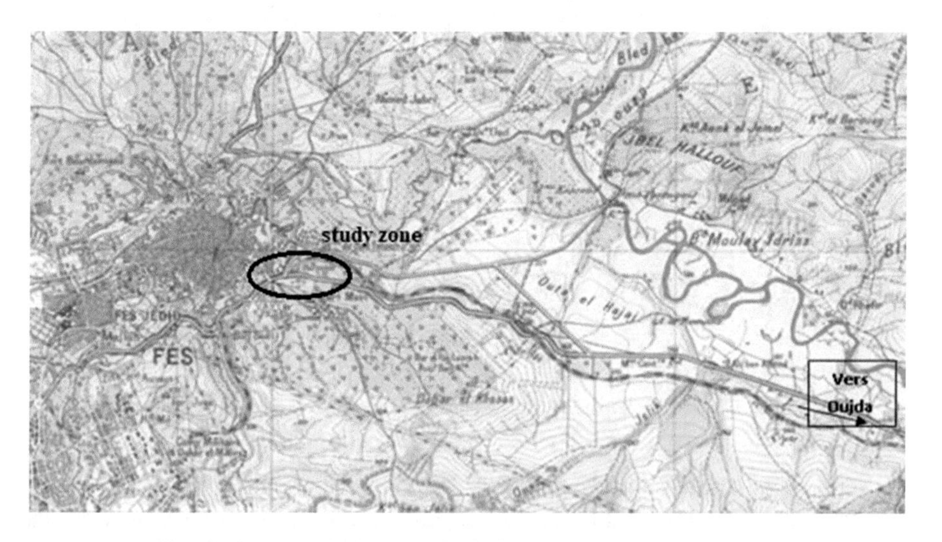

Fig. 4. Geographical and geological location of the embankment

A geotechnical companion consisting of the realization of soundings with samples taken intact, and the establishment of various devices: piezometric test, inclinometer test, Core drilling coupled with pressuremeter tests.

Soil samples are submitted, in the laboratory, to three groups of identification tests (granulometry, wet specific gravity, Atterberg limits).

That geotechnical investigations make it possible to distinguish in the embankment the following formations:

- Up to 12 m deep: the ground is loose marl.
- From 12 m to 22 m: We witness the passage to a more consistent yellow marl.
- From 22 m to 28 m: Where the marl becomes grayish and consistent considered as substratum.

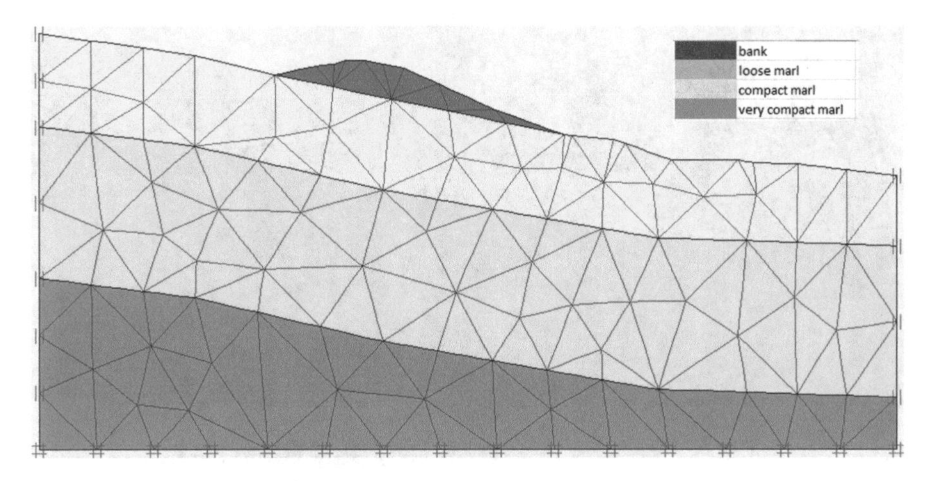

Fig. 5. Finite element model of the studied embankment

The results of the laboratory tests show that it is a marly soil, composed of 97% of fine elements, with a plasticity index Ip = 35%, which gives it the characteristics of a very clay soil. plastic that has a high sensitivity to water, so:

If it is mixed with increasing amounts of water, it eventually turns into mud. Clay has a liquid behavior.

On the contrary, if the clay is sufficiently dried out, the grains are very narrow and the bonds become intense. Clay has a strong behavior.

The behavior of the partially saturated soil resting on a substratum made of greyish marl was modeled with the elastoplastic law of Mohr-Coulomb.

A transient hydro-mechanical analysis was done using the Plaxis software 8.2 and the variations of the hydraulic variables, stresses and deformations were plotted. The analysis of the displacements obtained also made it possible to compare with the results of the inclinometers (Fig. 9) installed on both sides of the embankment.

Figure 6 shows the contours of the degree of saturation of the soil water as well as the deformations for the area of interest (Fig. 7). The orange areas in Fig. 6 indicate that the soil has been almost saturated. At the same time, in Fig. 7, it is noted that the shear stresses are mainly located along the interface between the greyish marl substratum and the loose marl soil layer, indicating that a fracture mechanism is produced. It can be seen that the shear zone merges with the zone affected by the infiltration of water.

Figure 8 also shows the calculated horizontal displacements, which are compared with the results of the installed inclinometers (Fig. 9). The maximum displacement obtained with the finite element calculation is equal to the displacement obtained with the inclinometers. We obtained a displacement of 18 mm.

Before the inflow of water, the compaction of the embankment calculated is of the order of 17 mm.

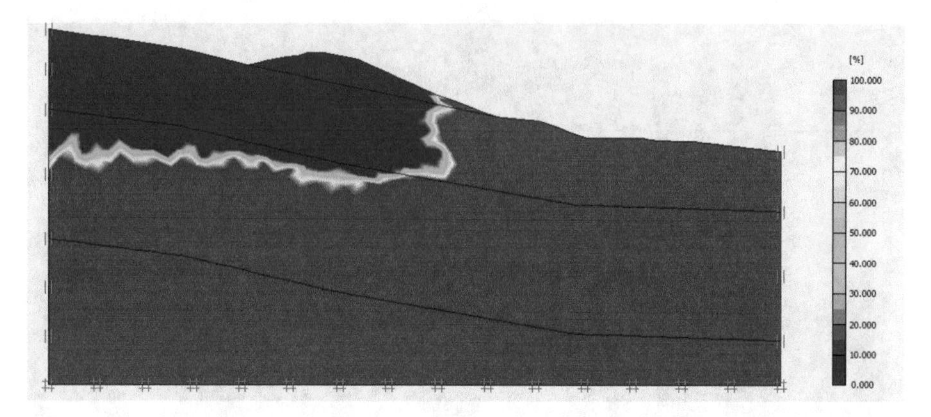

Fig. 6. Contours of the degree of saturation of the soil water

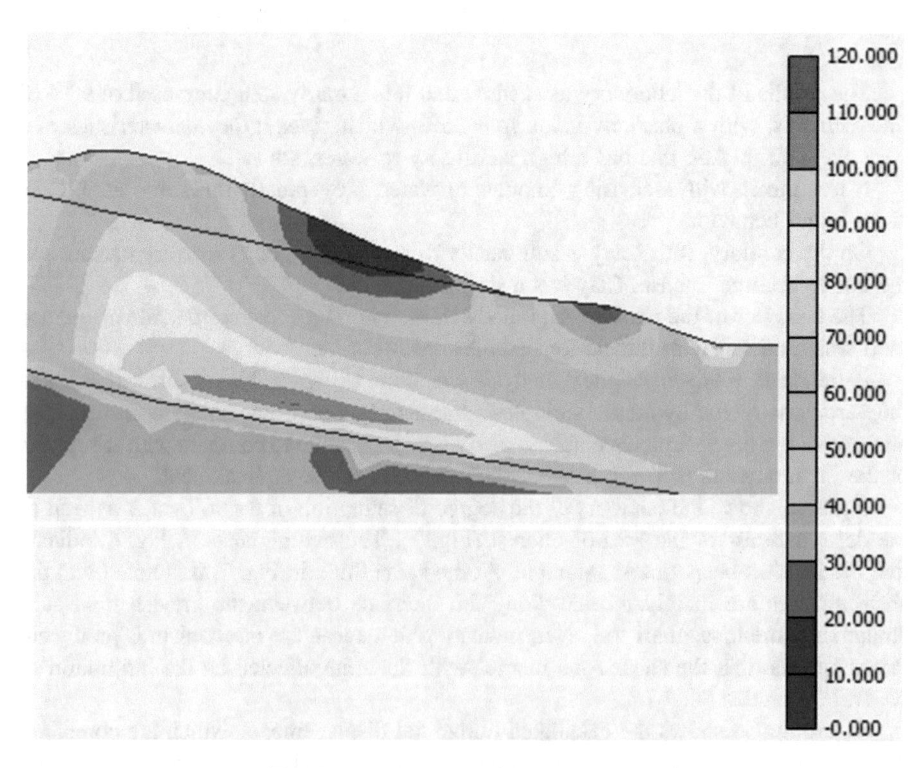

Fig. 7. Shear stresses along the slip line

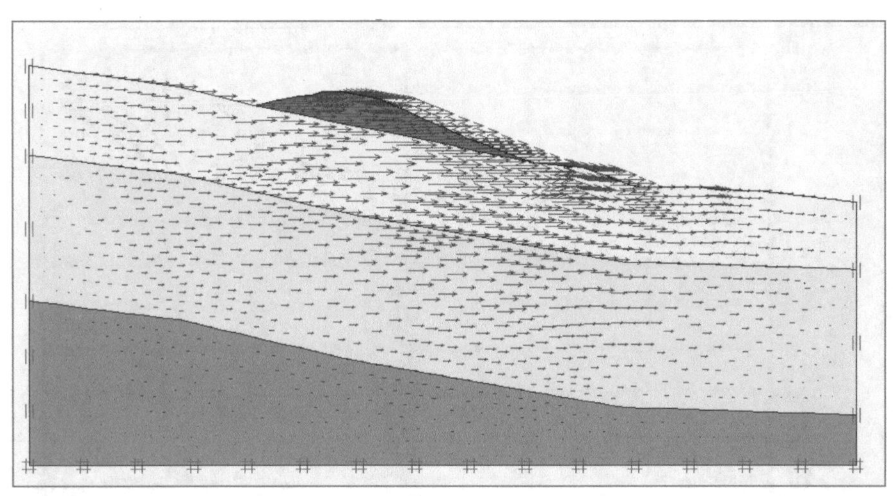

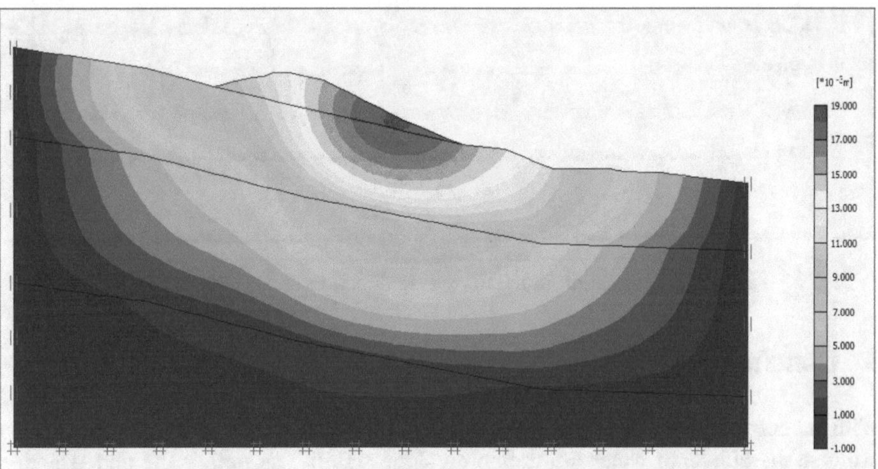

Fig. 8. Displacements obtained by Plaxis

But with the decrease of the mechanical characteristics of the soil of the embankment due to the inflow of water, we note that the settlement value becomes almost six times greater with Δh = 35 cm.

It is concluded that with the infiltration of water in the body of the embankment, the lift of the marly soil decreases and it becomes more and more deconsolidated, which justifies the important settlements noted.

The sharp increase in the degree of water saturation accompanied by large reductions in the volume of the solid skeleton qualitatively indicates the potential for rapid landslide.

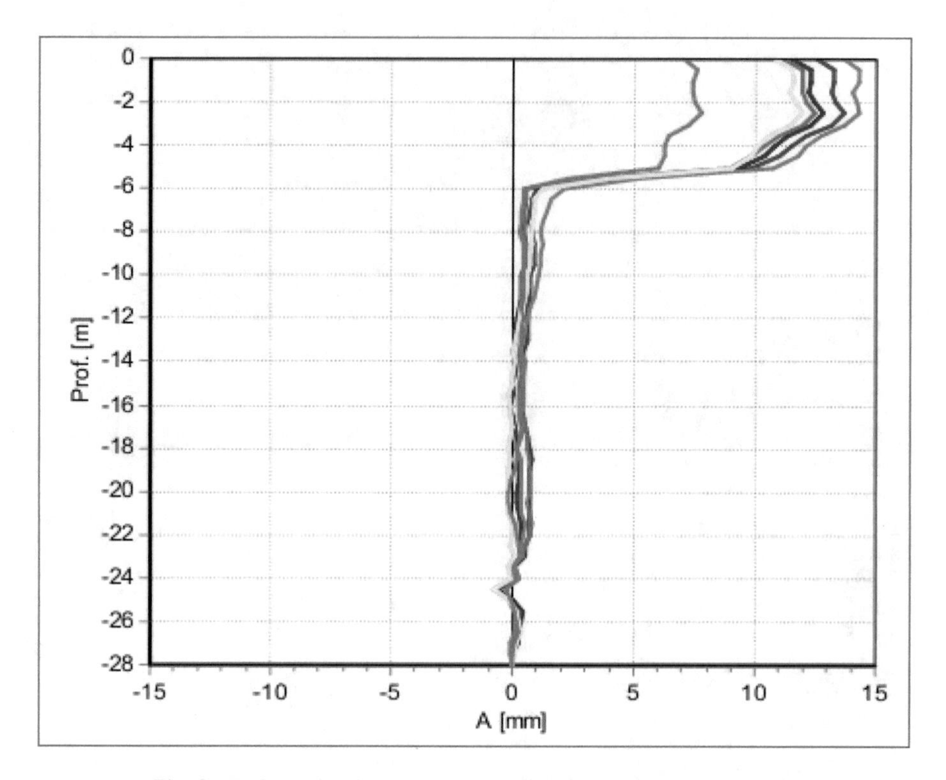

Fig. 9. Deformation in the transverse direction of the embankment.

5 Conclusions

With recent advances in numerical modeling of soil behavior, it is now possible to evaluate the effects of water infiltration on slope stability as well as the understanding of the key mechanisms in the onset and propagation of landslides.

Using the stress analysis of the unstable mass, detailed deformation makes it possible to draw interesting conclusions about the type of failure mechanism and to some extent about the behavior of the unstable mass after failure (e.g. rigid sliding in blocks or mud casting).

The present study made it possible to compare on a real geometrical model the results of calculation of the displacement maximal by different methods: inclinimeters installed on site and methods of the finite elements. The results obtained with both methods are similar.

Moreover, it is concluded that with the infiltration of water in the body of the embankment, the lift of the marly soil decreases and it becomes more and more deconsolidated, which justifies the important settlements noted.

The large increase in the degree of water saturation accompanied by strong reductions in the volume of the solid skeleton qualitatively indicates the potential for rapid landslide.

References

1. Eichenberger J., Nuth M., Laloui L.: Modeling landslides in partially saturated slopes subjected to rainfall infiltration. In: Laloui, L. (ed.) Mechanics of Unsaturated Geomaterials, pp. 235–250. Wiley (2010)
2. Eichenberger, J., Ferrari, A., Laloui, L.: Analyses géo-mécaniques des glissements de terrain superficiels. Mémoire de la Société vaudoise des Sciences naturelles **25**, 279 (2013)
3. Lu, N., Likos, W.J.: Unsaturated Soil Mechanics, p. 556. Wiley (2004). Mayoraz, P.F., Vulliet, L.: Neural networks for slope movement prediction. Int. J. Geomech. **2**(2), 153–173 (2002)

Features of Investigation of Soil According to Kazakhstan Norm and International Standards

A. S. Tulebekova[1(✉)], A. Zh. Zhussupbekov[1], T. Mussabayev[2], and S. Mussina[2]

[1] Department of Civil Engineering, L.N. Gumilyov Eurasian National University, Astana, Kazakhstan
krasavka5@mail.ru
[2] Eurasian Technological Institute, L.N. Gumilyov Eurasian National University, Astana, Kazakhstan

Abstract. Paper presented methodic of investigation of soil regarding requirement of Eurocode 7 and Kazakhstan standard. The methodic of investigation and testing of soil by these standards have some differences as long as Kazakhstan Standard is out of date and has not changed science 1994. Eurocode 7 is taking latest technology developments, technical methods and provides for the use of more modern equipment. Discussion of using geotechnical equipment's, technological features, advantages and disadvantages of aforementioned standards might be important for understanding and elimination of existing differences, for harmonization with international standards. Paper includes recommendation for the future modernization of Kazakhstan standards.

1 Introduction

Nowadays many international projects are realized in Kazakhstan. For example Italian and American companies which are work on western part of Kazakhstan met some complexity in non-correspondences of Kazakhstan codes to international. Kazakhstan engineers also feel some complexity during design modern projects where generally used advanced technologies.

Unfortunately, present Codes are confined application of modern technology of pile foundation installation, indicating incomplete usage of advanced technology such as CFA (continuous flight auger), DDS (drilling displacement system) or FDP (full displacement pile), Jet-grouting DMM (Deep mixing method) technologies and so on.

Aforementioned demands to using international Code, moreover, for realization unique project is required using leading foreign high-tech, economic, ecological and energy-efficient technology, including technology for pile installation, equipment for geological investigation, as well as laboratory testing.

Eurocode 7 - Geotechnical Engineering which was established in 2004 seems to be more reliable for adaptation for Kazakhstan construction condition. Many countries successfully accepted Eurocode 7 and during last years this Code becoming more international. Eurocode 7 is already showed itself as very elaborate design code where

S. Hemeda and M. Bouassida (Eds.): GeoMEast 2018, SUCI, pp. 142–148, 2019.
https://doi.org/10.1007/978-3-030-01941-9_12

given recommendations and requirements for most part of geoengineering process. It also allows using common international geoengineering terms and provides understanding between for designers, testing specialists, geotechnical engineers all over the World. Eurocode 7 includes recommendations and requirements for modern advanced technologies and embraced many aspects of modern geoengineering design.

The results of research is directed to developing of recommendation for modernization of Kazakhstan Codes and oriented to adaptation of advanced geotechnologies. The modernization will allow to complete use of advanced technologies capabilities in existing construction condition of Kazakhstan. Is also allow greatly reduce the expenses of the null cycle, which forming significant part (20%) of the total construction project expense. Development of recommendation for modernization of national geotechnical codes, oriented on adaption of advanced geotechnologies for piles installations in problematical soil conditions of Kazakhstan is very important for designers, testing specialists, geotechnical engineers as long as Codes of many countries has some differences due to of specific regional soil condition, and local specifications. Developing recommendation will be first stride for Kazakhstan construction to be of international part, to be understandable by world Geotechnics.

2 Eurocode 7

The development of Eurocode 7 has been strongly linked to the development of En 1990: Eurocode: Basis of structural design (CEN 2002) and the format for verifying ground-structure interaction problems is, of course common to both documents.

After giving the main contents of Eurocode 7, this contribution summarises the requirements relevant to pile design (without recalling the principles of LSD and of the partial factor method).

Design examples of piles under vertical compressive loadings can be found, for instance, in the proceedings of the workshop of ERTC 10 on the evaluation of Eurocode 7.

General rules is a rather general documents giving only the principles for geotechnical design inside the general framework of LSD. These principles are relevant to the calculation of the geotechnical actions on the structural elements in contact with the ground (footings, piles, basement walls), as well as to the deformations and resistances of the ground submitted to the actions from the structures. Some detailed design rules or calculation models, i.e. precise formulae or charts are only given in informative Annexes.

The Section on field tests in soil and rock includes cone and piezocone penetration tests CPT (U), Pressuremeter test PMT, flexible dilatometer test (rock and soil) FDT, code penetration test SPT, dynamic probing tests DP, weight sounding test WST, field vane test FVT, flat dilatometer test DMT and plate loading test PLT.

The Section on laboratory testing of soils and rocks deals with the preparation of soil and rock specimens for testing, tests for classification, identification and description of soil, chemical testing of soil and groundwater, strength index testing of soil, compressibility and deformation testing of soil, compaction testing of soil, permeability testing of soil, strength testing of soil, compressibility and deformation testing of soil, tests for classification of rocks, swelling testing of rock material and strength testing of rock material.

It also includes a number of informative Annexes with examples of correlations and derivations of values of geotechnical parameters from field test results. The informative Annexes D.6 & D.7 for CPT tests, and E.3, for PMT tests, are such examples for determining the compressive resistance of a single pile.

The core of Section 7 of EN 1997-1 is devoted to the behavior of pile foundations under axial (vertical) loads. The importance of static load tests is clearly recognized as the basis of pile design methods. An innovative concept introduced in this section, with regard to traditional pile design, is the use of correlation factors ξ for deriving the characteristic compressive and tensile resistances, of piles either from static pile load tests or from ground test results. In both cases, the correlation factors ξ depends mainly on the number of tests performed, whether pile load tests or profiles of ground tests.

3 Comparison of Eurocode with Kazakhstan Codes

Nowadays is put into action recommendation of Eurocode 1–8 which are replacing National Codes. However some Countries arc transforming it's National Codes by taking into account Eurocode recommendation. The principal deference between Eurocode and Kazakhstan Code (SNiP RK) is absence of requirements for the geotechnical design in Kazakhstan Codes. In Eurocode the strategy of geotechnical design includes interaction of two researches - geological engineering and geotechnical. However today is difficult to design without qualitative geotechnical investigation. Geotechnical research include results of engineering and geological investigation which are had been used during definition of soil and foundation modeling. Recommendation of Eurocode promote to mutual researchers and designers work. Unfortunately in Kazakhstan practice engineering and geological investigation is one different part of the design, and frequently no interaction between researchers and designers. The program of the geological investigation rarely coordinate to designer and as a result there are absent common strategy of the design.

One another difference of Eurocode is design procedure. According to Kazakhstan Code the design of soil basement is recommended to carry out by three steps. During the first and second step of foundation design it is allowable to use preliminary strengthen and deformative properties of soil taken from table of SNiP RK, during the last third step it is required to perform both laboratory and field test to approve design project. According to Eurocode for all of this three engineering and geological investigation steps for definition strengthen and defomative properties of soil is required to use results of laboratory or field tests only. Moreover Eurocode use term « derive value » that mean value of geotechnical parameter of soil obtained by results of identical laboratory or field test of soil by using correlation relationship or using inverse calculation. For example deformation modulus obtained by independence tests: laboratory tests, field tests by dilatometer, by correlation relationship with physical parameter, by results of well known settlement calculation (Boldyrev et al. 2010).

Laboratory tests which recommended by Eurocode are presented in Table 1. It is well known that the obtained by laboratory tests mechanical and physical properties of soil depends on quality of soil sample. Eurocode differ five categories of soil samples quality assuming that the properties of soil invariable due to sampling, packing and transportation.

Table 1. Eurocode soil laboratory methods

Parameter	Type of soil		
	Gravel	Sand	Sandy loam
Compression deformation Modulus (E_{oed}) Compression index (c_c)	OED, TX	OED, TX	OED, TX
Elastic Modulus (E) Shear Modulus (G)	TX	TX	TX
Consolidated drained strength parameters (φ, c)	TX, SB	TX, SB	TX, SB
Residual strength (φ_R, c_R)	RS, SB	RS, SB	RS, SB
Undrained strength, (c_u)	–	–	TX, DSS, SIT
Density of soil (p)	BDD	BDD	BDD
Consolidation (c_N)	–	–	OED, TX
Filtration	TXCH, PSA	TXCH, PSA	PTS, TXCH, PTF
Parameter	Type of soil		
	NC clay	OC clay	Silt
Compression deformation Modulus (E_{oed}) Compression index (c_c)	OED, TX	OED, TX	OED, TX
Elastic Modulus (E) Shear Modulus (G)	TX	TX	TX
Consolidated drained strength parameters (φ, c)	TX, SB	TX, SB	TX, SB
Residual strength (φ_R, c_R)	RS, SB	RS, SB	RS, SB
Undrained strength, (c_u)	TX, DSS, SB, SIT	TX, DSS, SB, SIT	TX, DSS, SB, SIT
Density of soil (p)	BDD	BDD	BDD
Consolidation (c_N)	–	–	OED, TX

According to Eurocode swelling of soil research for rock soil, whereas according to Kazakhstan Codes concern to unstable structure.

Comparison of national Codes with Eurocode (Boldyrev et al. 2010) shown that there are not yet developed recommendation for performing following field tests:

- cone penetration test with analysis of pore water pressure (CPTU).
- cone penetration lest by dynamic load (SPT).
- dilatometer test.

However in Eurocode is absent some recommendation for field tests in condition of frozen soil (Boldyrev et al. 2010).

Modern megaprojects put forward modern requirements to engineers. This led to refuse from traditional out-of-dates technologies (traditional boring and driving diesel-hammer piles) and use new more economical and reliable technologies like CFA (continuous flight auger), DDS (drilling displacement system), steel "H" piles.

Another advantage of Eurocode is recommendation for design and calculation of modern pile technologies such as CFA, FDP and so on. Kazakhstan Code has not some recommendation for these piles technology, and so designers have to use recommendation for traditional out-of- dates pile technologies (traditional boring and driving diesel-hammer piles). As a result incomplete usage of modern technology has a place. Design of modem pile by Kazakhstan Code is not include many technologies factors such as high value of concrete pressure during CFA pile installation and soil displacement without excavation during DDS pile installation.

Table 2. Safety factor of pile tests comparison

Code	Safety factor			Number of test required
	Design	SLT	DLT	
Eurocode		2.18		If number of tests equal or less than 2
		1.91		If number of tests equal or greater than 20
			2.23	If 2 DLTs is performed
			1.95	If number of DLTs greater than 20
Kazakhstan code	1.5	1.4		SLTs on 0.5% of constructed piles on construction site (2 SLTs at least in a site)
			1.2	At least 6 DLTs in a site (or 1% of working piles on construction site)

Comparison of safety factors recommended by Kazakhstan Code and Eurocode is presented in Table 2. In this table are listed design safety factor, safety factor for static load test, and safety factor for dynamic load test, together with the number of tests required or specified for a pile construction site.

In Kazakhstan is not equipment for testing some types of field tests of soil, which showed in Table 3.

Table 3. Field test methods by Eurocode and Kazakhstan Codes

Field test (Eurocode 7)	Codes	SNiP, GOST analogy	Equipment which used regarding traditional norms (Kazakhstan)	Results of tests
Cone penetration test (CPT)	EN ISO 22476-12	GOST 19912-2001 SNiP RK 5.01-03-2002	Static sounding, cone with measure q_c, f_c, R_f	– resistivity of soil under cone – resistivity of soil on lateral surface – friction coefficient
Piezocone penetration tests (CPTU)	EN ISO 22476-1	–	No equipment for measure pore water pressure	– resistivity of soil under cone – resistivity of soil on lateral surface – neutral stress
Dynamic probing test (DPT)	EN ISO 22476-2	GOST 19912-2001 SNiP RK 5.01-03-2002	Dynamic sounding, probe for measuring the number of impact	– number of impact N_{10} for tests: DPL, DPM, DPH – number of impact N_{10} or N_{20} for DPSH test
Standard penetration test (SPT)	EN ISO 22476-3	–	Dynamic sounding, probe for measuring the number of impact	– number of impact N – impact energy
Cone penetration test (CPT)	EN ISO 22476-12	GOST 19912-2001 SNiP RK 5.01-03-2002	Static sounding, cone with measure q_c, f_c, R_f	– resistivity of soil under cone – resistivity of soil on lateral surface – friction coefficient
Pressiometer test	EN ISO 22476-4	GOST 20276-85	No equipment	– pressiometer module of deformation E_M – slippage pressure pf – limiting pressure plm – graphic of measuring
Field test (Eurocode 7)	Codes	SNiP, GOST analogy	Equipment which used regarding traditional norms (Kazakhstan)	Results of tests
Weight sounding test (WST)	EN ISO 22476-10	–	No equipment	– record the time variation of -resistivity sounding
Field vane test (FVT)	EN ISO 22476-9	GOST 21719-80	No equipment	– undrained shear resistance – graphic of measuring – specific cohesion cb
Pressiometer test	EN ISO 22476-4	GOST 20276-85	No equipment	– pressiometer modul of deformation E_M – slippage pressure pf – limiting pressure plm – graphic of measuring
Weight sounding test (WST)	EN ISO 22476-10	–	No equipment	– record the time variation of -resistivity sounding

4 Conclusions

Eurocode 7 - Geotechnical Engineering which was established in 2004 seems to be more reliable for adaptation for Kazakhstan construction condition. Many countries successfully accepted Eurocode 7. Eurocode 7 is already show itself as very elaborates design code where given recommendations and requirements for most part of geo-engineering process. It also allows using common international geoengineering terms and provides understanding between for designers, testing specialists, geotechnical engineers all over the World. Eurocode 7 includes recommendations and requirements for modern advanced technologies and embraced many aspects of modern geoengineering design.

By result of Kazakhstan Code comparison with Eurocode it is become obvious one disadvantage of Kazakhstan Code is absence of recommendation for design, testing and calculation modern pile technologies such as CFA, DDS and so on. Eurocode is also presented by unified documentation for geoengineering comparing with Kazakhstan Code, where many Codes are and where someone may contradict to other.

Comparison also shown that in Kazakhstan Codes is not developed recommendation for performing following field tests: cone penetration test with analysis of pore water pressure, cone penetration test by dynamic load and dilatometer tests.

References

EN 1997-1:2004.: Eurocode 7. Geotechnical Design. Part I. General rules

EN 1997-2:2007.: Eurocode 7. Geotechnical Design. Part 2. Design assisted by laboratory and field testing

SNiP RK 5.01-03-2002 - Pile foundation. Astana, KAZGOR (2003)

Boldyrev, G.G., Idrisov, I.Kh., Barvashov, V.A.: Comparison of Eurocode 7. Part 2 with Russian Codes. Similarities and Differences. Technical Regulation. Engineering Investigation, pp. 22–26 (2010)

MSP 5.01-101-2003 - Design and installation of pile foundation

Marsh Funnel Test as an Alternative for Determining the Liquid Limit of Soils

Taha M. Khalaff(✉) and Adam Lobbestael

Department of Civil and Architectural Engineering, Lawrence Technological University, 21000 West Ten Mile Road, Southfield, MI 48075, USA
tkhalaff@ltu.edu

Abstract. The liquid limit of a soil is a fundamental index property that is routinely used in engineering practice to predict the behavior of cohesive soils. The Casagrande and fall cone test methods are the standard methods used to determine liquid limit. However, these methods require significant sample preparation, are time-intensive, and are subject to a range of errors dependent on the experimenter's technique and experience. The study presented here investigates the feasibility of an alternative method for determining the liquid limit of soil. The nature of the definition of liquid limit suggests a relationship between viscosity, a mechanical fluid property, and the liquid limit. Accordingly, this study investigates the relationship between liquid limit, determined by Casagrande method, and the Marsh Funnel viscosity. The Marsh Funnel viscosity test produces an indicator of absolute viscosity by observing the time required for a fluid to flow through a calibrated funnel. Four different types of soil were prepared at a range of successively decreasing moisture contents. The liquid limit of the material was taken as the moisture content at which the Marsh Funnel viscosity approached infinity (stopped flowing). This value was then compared to the liquid limit determined using the Casagrande method in order to make preliminary conclusions regarding the suitability of using the Marsh Funnel test as an alternative method to determine liquid limit.

1 Introduction

The liquid limit is one of the index properties of cohesive soils which is used extensively by geotechnical engineers. It is used for the classification of fine grained soils, as well as a correlative parameter in preliminary estimation of many physical and engineering properties. There are two methods at present to determine liquid limit of soils: (1) percussion method and (2) cone penetration method. The percussion method (or Casagrande's falling cup method) is included in ASTM standards (ASTM Designation D4318) and is still used in much of the world. Johnston and Strohm (1968), Wroth and Wood (1978), Whyte (1982), Lee and Freeman (2007), Kayabali and Tufenkci (2010), and Haigh (2012), determined many limitations and uncertainties in the Casagrande method. These limitations and uncertainties included material, dimensions and weight of the cup, soil type, frequency of drops, the tendency of the halves to slide together, the migration of water in dilatant soils, and operator judgment for closure length of the groove.

S. Hemeda and M. Bouassida (Eds.): GeoMEast 2018, SUCI, pp. 149–159, 2019.
https://doi.org/10.1007/978-3-030-01941-9_13

Because of these limitations and uncertainties involved in the Casagrande test method, researchers have attempted to find better and more reliable alternative test method that determines liquid limit of cohesive soils. In 1955, Darienzo and Vey used a vane apparatus (small vane) to correlate the vane moment with both plastic and liquid limits. Ramachandran et al. (1963) developed an equation that correlated the liquid limit with the amount of dye adsorbed by clayey soil. By using two different dye adsorption solutions, malachite green and methylene blue, Ramachandran et al. found that is a linear relationship between liquid limit and amount of dye adsorbed. Russel and Mickle (1970) found linear relationships between consistency limits and moisture content that were obtained at specific pressure intensity by using the soil moisture tension method, which made it possible to correlate the consistency limits with moisture tension. Prakash and Sridharan (2002) used the equilibrium sediment volume method to develop formulas that correlated liquid limits determined by Casagrande and cone penetration methods with sediment volume. Nagaraj and Sridharan (2010) found linear relationship between liquid limit determined by cone penetration method and k_o - Stress method when the sample under k_o-stress of 0.9 kPa. This linear relationship made it possible to determine liquid limit by using vertical consolidation stress method. Many other research efforts were made by various researchers to develop better and more reliable methods to determine liquid limit.

In the early 1900s, Atterberg defined the liquid limit as the water content of soil where the behavior of a cohesive soil changes from plastic to liquid or from liquid to plastic. Also, some researchers such as Varnes (1978), Cruden and Varnes (1996) and Hungr et al. (2001), found that during mudflow movement from the initiation of the movement to the actual movement, the soil mass could change rapidly from a plastic state to a viscous liquid state, which concluded that the mudflow moisture content closely related to the liquid limit of cohesive soil. Based on the Atterberg liquid limit definition (Varnes 1978; Cruden and Varnes 1996; Hungr et al. 2001) conclusions, and the viscosity as one of the mechanical properties of liquids, there is some merit exploring how the liquid limit may be obtained by using apparatus that used to determine mudflow viscosity. In this research, Marsh Funnel was used to determine the relative viscosity of mudflow. This new approach is superior to the conventional methods in several aspects as follows:

1. It does not require experience, the operator dependency is no longer a matter of concern.
2. The test duration is remarkably short. All the data needed to find the liquid limit is obtained in about 30–60 min.
3. The device is simple and lightweight. It can be used anywhere in a laboratory.

In this paper a comparative study is conducted on four different types of soils, two of them being essentially Kaolinite, one Montmorillonite, and one natural soil (possibly Illite), by using the Casagrande method along with the Marsh Funnel Viscosity method.

2 Material Selection and Proposed Liquid Limit Test Method

Materials

Kaolinite, montmorillonite, and illite, according to the plasticity chart for mineralogical analysis published by Mitchell (1993), are types of clay minerals that behave quite differently from each other under any given set of physical-chemical environments. Hence, the mechanisms governing the liquid limit of all three types of soils may be the same, but the resulting behavior may be different. This investigation was conducted on four different types of soils, one being a natural soil and three being processed materials. The percussion method liquid limit (LL) values of these soils ranged from 28 to 64%, while the plasticity index (PI) values ranged from 10 to 39%. The physical properties of the tested soil, according to ASTM Standards are summarized in Table 1

Table 1. Physical properties of soil used in the present study

Soil no	Soil type	LL (%)	PL (%)	PI (%)	Grain size distribution			
					Gravel (%)	Sand (%)	Silt (%)	Clay (%)
1	100% K	62	33	29	0	0	0	100
2	20% K, 40%M, 40% S	28	18	10	0	40	40	20
3	10% B, 45%M, 45% S	61	22	39	0	45	45	10
4	Natural Soil (SP-SC)	43	20	23	0.61	80.32	6.2	12.9

B: Bentonite; K: Kaolinite; M: Silt; S: Sand

and grain size distribution shown in Fig. 1.

While the plasticity ranges in this research appear to cover types of soils found in nature, it was important to ensure that the method proposed is applicable to a variety of soils and thus artificial soils prepared in the laboratory were employed. To achieve this goal, kaolinite, bentonite, fine sand, and silt were used as additives. To obtain soils with higher plasticity, one of the most abundant soil samples in the laboratory was mixed with bentonite, with varying ratios of dry mass. In this way, soils with a liquid limit of up to 64 were obtained. Similarly, one of the soil samples was mixed with different amounts of fine sand to obtain soil samples with liquid limits below 28. At the end, 4 soil samples were prepared. All soil samples, were sieved through #40 mesh prior to conducting the conventional and newly proposed methods of liquid limit tests. In addition, these soils were chosen due to the differences of clay mineralogy (Kaolinite, Illite, and Montmorillonite), and their liquid limit governing mechanism. According to (Sridharan et al. 1986, 1988) the kaolinite soil governing mechanisms of liquid limit is controlled by the mode of particle arrangement, as determined by the inter-particle forces, but that of montmorillonite soil and possibly illite is due primarily to double-layer held water.

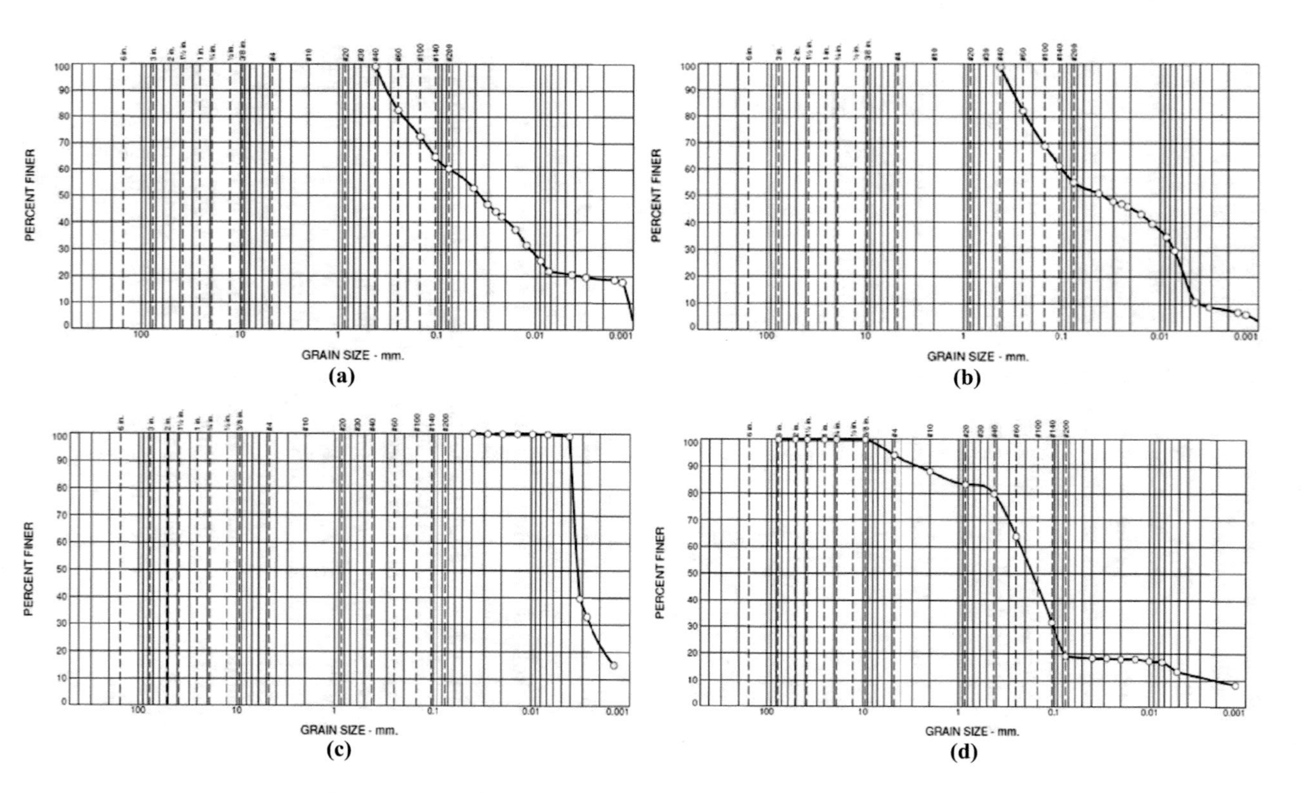

Fig. 1. Grain size distribution curves of: (a) Mixed soil (10% B, 45% M, 45% S), (b) Mixed soil (20% K, 40% M, 40% S), (c) Pure Kaolinite, (d) Natural soil (SP-SC)

Proposed Test Procedure

The apparatus used in this investigation essentially consisted of a commercially available Marsh Funnel (Figs. 2 and 3). The Marsh Funnel viscosity ASTM D6910, is reported as seconds and used as an indicator of the relative consistency of fluids. ASTM D6910-04 is updated in 2009 to ASTM D6910/D6910 M and there is no changes in the test procedure or the funnel cone size. ASTM D6910/D6910 M requires that either the values stated in SI units or English units must be regarded separately as standard. Because the values stated in each system may not be exact equivalents; therefore, each system shall be used independently of the other. Combining values from the two systems may result in nonconformity with the standard.

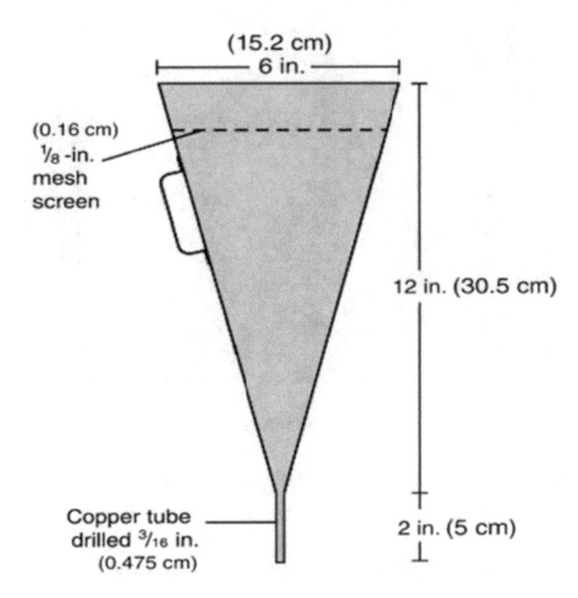

Fig. 2. Standard marsh funnel (Balhoff et al. 2011).

The liquid limit (LL) and plastic limit (PL) for each sample were determined in accordance with ASTM D 4318. Four replicated tests were performed on each soil used in this study. The liquid limit results were used to compare them with the proposed method results.

As in the Marsh Funnel test, each soil sample of 1000 g weight (consisting of the - #40 sieve fraction) was put into a plastic jar. A sufficient amount of distilled water was added and carefully mixed with a spatula until the point where the soil mass could be slowly poured out of the jar (Fig. 4(a)). Care was taken that the soil was not so wet as to have free water on the surface when standing. The samples were allowed to stand in the capped jars for 16–24 h at room temperature 27° ± 2 °C.

Immediately before running the test, distilled water was added to the sample until saturation which gave a water content higher than the percussion LL. Having water content higher than the percussion LL would lead to a viscous liquid state (mudflow).

Fig. 3. Marsh funnel apparatus used in this research

The mudflow soil was placed into an electric mixer stirring vigorously for 10 min, to help it reach a homogenous state (Fig. 4(b)). Upon reaching equilibrium, about 2.9 L (3 quarts) was obtained. Then, the fluid was placed in the Marsh Funnel. The Marsh Funnel is designed so that 1500 mL of fluid can be poured into the funnel. A small stopper is placed in the orifice at the bottom to prevent flow out while the fluid is poured into the funnel (Fig. 4(c)). Once it reaches 946 mL, a beaker is placed in position below the funnel. The plug is removed from the orifice and the fluid is allowed to flow until the fluid completely drains (Fig. 4(d)). The flow time (t) is recorded, and then the density (ρ) and water content (w) are determined. Each type of material was tested 5 times with different moisture contents. All tests were repeated 3 times at same range of moisture content and good repeatability was obtained.

Marsh Funnel Viscosity was calculated by using the following Eq. (1)

$$\mathrm{MFV} = \rho(\mathrm{t}-25) \tag{1}$$

Where MFV is Marsh Funnel Viscosity ($\mathrm{g/cm^3.sec}$); ρ is slurry density ($\mathrm{g/cm^3}$); t is the time (sec) that takes 946 ml (1 quart) to flow through Marsh Funnel.

Other than requiring that all tests on a mud be made at the same temperature, no temperature control was exercised. The density of each sample was determined by weighing a known volume, and moisture content was determined for each single test.

Fig. 4. Marsh funnel test: (a) Saturated sample mixed with a spatula (b) Prepared mudflow sample, (c) Mudflow poured into the funnel, (d) Mudflow allowed to flow.

3 Results and Discussion

Figure 5 shows the Marsh Funnel Viscosity versus water content, obtained in the Marsh Funnel test for four different soil types. The Marsh Funnel Viscosity (MFV), of the four different soils identified in Table 2, was determined at various water contents (w). All water contents exceeded the Casagrande liquid limit of the soil. Based on the slope and the curve type, the first degree of rational trend-line was assumed to be more reliable to determine proposed liquid limit (LL_p). The data can be fitted quite well (regression coefficient ranges from 78% to 98%) by rational function expressed as

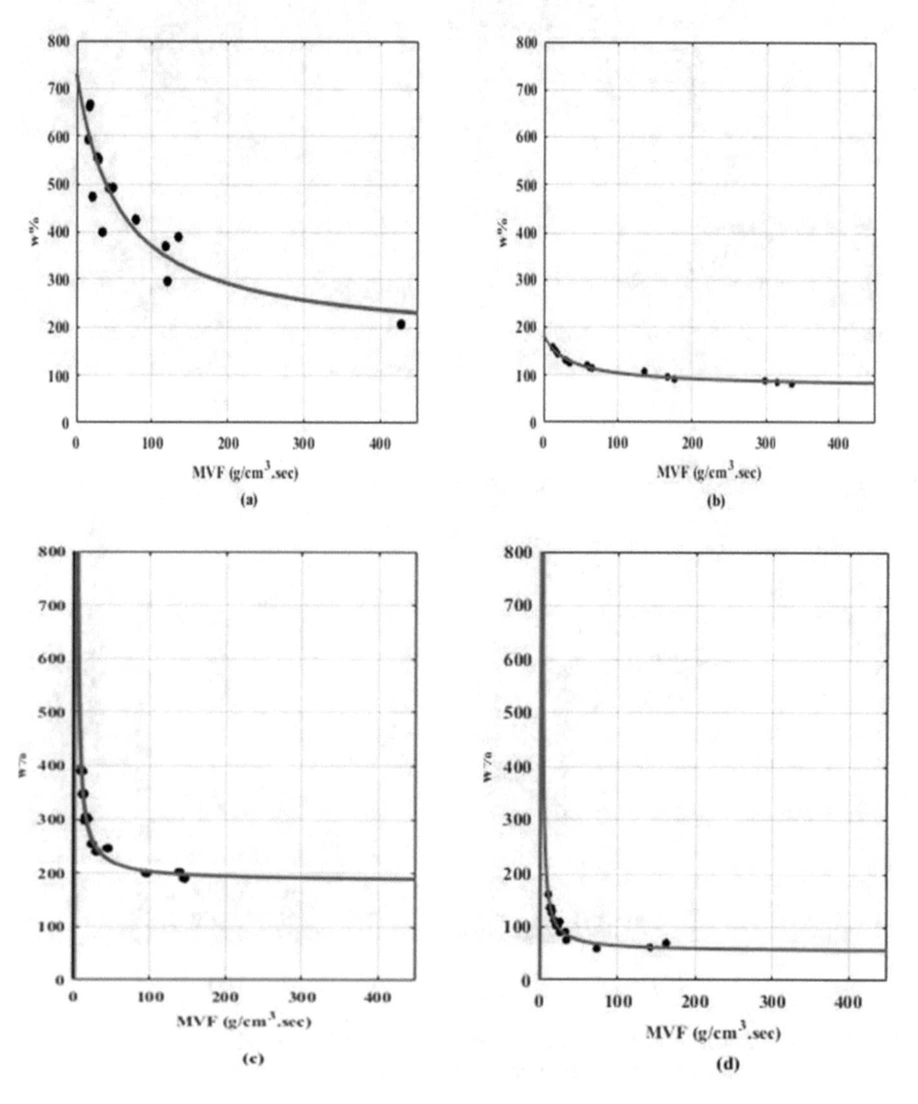

Fig. 5. Marsh Funnel Viscosity (MFV) vs. Water content (wc): (a) Mixed soil (10% B, 45% M, 45% S), (b) Mixed soil (20% K, 40% M, 40% S), (c) Pure Kaolinite, (d) Natural soil (SP-SC)

Table 2. Proposed liquid limit test results compared with Casagrande test results.

Soil material	Liquid limit	
	Casagrande method	Marsh funnel method
Pure Kaolinite	62	186
20%K, 40%M, 40%S	28	77
10%B, 45%M,45%S	61	169
Natural Soil (SP-SC)	43	56

$$LLp = \lim_{MFV \to \infty} \frac{a.MFV + b}{MFV + c} \quad (2)$$

a, b, and c are constants

The degree of reduction in MFV was different for each water content because the sample density is a function of the MFV. The results confirm that viscosity affected by increases in the water content (Kooistra et al. 1998). As seen in Fig. 5, the repeated test results at same range of water content has almost the same Marsh Funnel viscosity. Each type of soil tested and analyzed by both methods, the Casagrande and the Marsh Funnel method, for comparison of possible relations between them. The results listed in Table 2. Analyzing the results obtained by the Marsh Funnel method can be observed that for the tested soil, the liquid limit would be higher than that determined by Casagrande method, approximately three times.

Natural soil (SP-SC) has low plasticity when compared with other soils because the soil contains possibly Illite and probably uniformly graded grain size distribution and average particle size (not proven, only assumed, and based on its workability).

Mixed Bentonite possesses a smaller amount of clay minerals and a larger amount of non-plastics and impurities, when compared with mixed Kaolinite. As they have almost similar size distribution, Mixed Bentonite is more plastic than the mixed Kaolinite, as shown in Fig. 5(a) and (b), respectively.

The Marsh Funnel method, which is the proposed method, based on the time to flow about 946 ml for each sample, shows differences among the plastic behavior of the studied minerals. Mixed Kaolinite is the less plastic mineral, an aspect not shown with the Marsh Funnel method. This result is confirmed by the mineralogical analysis, which shows a smaller amount of clay minerals for Natural Soil when compared with other tested soils, regardless of their particle size distribution. In the case of soils with mixed clay minerals, the relative values of the liquid limit obtained by the percussion and the Marsh Funnel methods depend upon the compensating effects due to montmorillonite and kaolinite.

As the proposed method shows, Bentonite is the most plastic mineral, because the force needed to deform it is the smallest; its mineralogical analysis can confirm its plastic behavior due the massive presence of clay minerals, in the form of Montmorillonite.

Finally, these observations indicate that the clay mineral type and its proportion in the clay content are responsible for Marsh Funnel method giving a higher value of liquid limit than the other method, but not the clay content alone.

4 Conclusions

The March Funnel tool is introduced as an alternative tool to determine the Atterberg limits on a more rational and quantifiable basis. The empirical equations based on the experimental data from four soil samples help determine the liquid limit with a degree of accuracy. Results from this experimental investigation done on four different soils, two of them being essentially kaolinite, one possibly illite, and one montmorillonite indicate that the liquid limit obtained by Marsh Funnel are observed to be three times higher than percussion method irrespective of the clay mineral type present in the soil.

In this method both the viscous shear resistance and frictional shear resistance seems to work simultaneously and depending on the type of clay mineral being present in the soil, that particular mechanism dominates and becomes the controlling mechanism.

Further speculation regarding the relationships between Marsh Funnel Viscosity and liquid limit cannot be made until further experiments are completed using a wider variety of soil types.

References

ASTM: Standard test methods for liquid limit, plastic limit, and plasticity index of soils. ASTM standard D4318-10. American Society for Testing and Materials, West Conshohocken, PA, USA (2010)

ASTM: Standard test method for Marsh Funnel viscosity of clay construction slurries. ASTM standard D6910-04. American Society for Testing and Materials, West Conshohocken, PA, USA (2004)

ASTM: Standard test method for Marsh Funnel viscosity of clay construction slurries. ASTM standard D6910D6910 M-09. American Society for Testing and Materials, West Conshohocken, PA, USA (2009)

Cruden, D.M. Varnes, D.J.: Landslide types and processes, landslides: investigation and mitigation. Transportation Research Board, pp. 36–75 (1996)

Darienzo, M., Vey, E.: Consistency limits of clays by the vane method. Proc. Highw. Res. Board Ann. Meeting. **34**, 559–566 (1955)

Haigh, S.K.: Mechanics of the Casagrande liquid limit test. Can. Geotech. J. **49**, 1015–1023 (2012)

Hungr, O., Evans, S.G., Bovis, M.J., Hutchinson, J.N.: A review of the classification of landslides of the flow type. Environ. Eng. Geosci. **7**(3), 221–238 (2001)

Johnston, M.M., Strohm, W.E.: Results of second division laboratory testing program on standard soil samples. Misc. Paper No. 3-978, U.S. Army Engineer Waterways Experiment Station, Vicksburg, MS (1968)

Kayabali, K., Tufenkci, O.O.: Determination of plastic and liquid limits using the reverse extrusion technique. Geotech. Test. J. **33**(1), 14–22 (2010)

Kooistra, A., Verhoef, P.N.W., Broere, W., Ngan-Tillard, D.J.M., van Tol, A.F.: Appraisal of stickiness of natural clays from laboratory tests. In: Proceedings of the National Symposium of Engineering Geology and Infrastructure, Delft, Netherlands, pp. 101–113 (1998)

Lee, L.T., Freeman, R.B.: An alternative test method for assessing consistency limits. Geotech. Test. J. **30**(4), 1–8 (2007)

Balhoff, M.T., et al.: Rheological and yield stress measurements of non-Newtonian fluids using a marsh funnel. J. Pet. Sci. Eng. **77**(11), 393–402 (2011)

Mitchell, J.K.: Fundamentals of Soil Behavior, 2nd edn. Wiley, Hoboken (1993). (Chap 3)

Nagaraj, H.B., Sridharan, A.: An improved ko-stress method of determining liquid limit of soils. Int. J. Geotech. Eng. **4**(4), 549–555 (2010)

Prakash, K., Sridharan, A.: Determination of liquid limit from equilibrium sediment volume. Géotechnique **52**(9), 693–696 (2002)

Ramachandran, V.S., Kacker, K.P., Rao, H.A.B.: Determination of liquid limit of soils by dye adsorption. Soil Sci. **95**, 414 (1963)

Russel, E.R., Mickle, J.L.: Liquid limit values by soil moisture tension. J. Soil Mech. Found. Div. Proc. ASCE **96**, 967–989 (1970)

Sridharan, A., Rao, S.M., Murthy, N.S.: Liquid limit of montmorillonite soils. Geotech. Test. J. **9** (3), 156–159 (1986)

Sridharan, A., Rao, S.M., Murthy, N.S.: Liquid limit of kaolinitic soils. Géotechnique **38**(2), 191–198 (1988)

Varnes, D.J.: Slope movement types and processes, landslides: analysis and control, transportation. Research Board, Washington, D.C., USA (1978)

Whyte, I.L.: Soil plasticity and strength e a new approach for using extrusion. Ground Eng. **15** (1), 16–24 (1982)

Wroth, C.P., Wood, D.M.: The correlation of index properties with some basic engineering properties of soils. Can. Geotech. J. **15**(2), 137–145 (1978)

Contribution to the Study of Geotechnical Characterization and Behaviour of Tunis Soft Clay

Nadia Mezni and Mounir Bouassida(✉)

Université de Tunis El Manar – Ecole Nationale D'Ingénieurs de Tunis LR14ES03-Ingénierie Géotechnique, BP 37 Le Belvédère, 1002 Tunis, Tunisia
mounir.bouassida@enit.utm.tn

Abstract. Tunis soft clay being known as one of the most problematic soils has poor mechanical characteristics, high compressibility and exhibits fragile shear strength. This paper considers the geotechnical characterization of Tunis soft clay by compiling results from in situ and laboratory tests. Accordingly, some correlations are suggested. The assessment of observed behavior of Tunis soft clays in the zone of interchange ramps was investigated. The follow up of ramps behavior was performed for a period of three months. The evolution of settlement was monitored by rod settlement, hydraulic settlement and multi-points settlement. A plane strain model was built for numerical investigation conducted by Plaxis software to simulate the behavior of the ramp's embankment. Hardening Soft Soil Model (HSM) and Soft Soil Model (SSM) were adopted for the soft clay layer. The results showed an agreement between the predictions of the two models of the behavior of the soft clay. Using measured settlement, the adopted behavior for Tunis soft clay is justified.

1 Introduction

Numerous research projects were carried out to characterize soft soils in the entire world. These researches recommend the suitable investigation tools for in-situ and laboratory testing and to propose useful correlations ensuring the best geotechnical characterization of soft clays. In this view different methods were suggested for the estimation of soft soils properties, in particular the determination of undrained cohesion, a key parameter for the design,

The ground of Tunis is mainly constituted by clay layers often very soft over the first 20 m depth. This soft soil causes significant problems especially in terms of geotechnical identification because of the extraction of undisturbed samples which is too difficult and also when performing the in situ tests. From several oedometer tests conducted on intact specimens extracted at various depths less than 20 m the over-consolidation ratio of Tunis soft clay was found in the range of 0.8–1.1 (Bouassida 2006). Although those results indicate that Tunis soft soil is classified as slightly under consolidated it is almost assumed normally consolidated soil.

Besides, the design of foundations on such soil requires some precautions. Among the well known cases history in Tunis, the construction of the 20 floors Africa hotel

S. Hemeda and M. Bouassida (Eds.): GeoMEast 2018, SUCI, pp. 160–174, 2019.
https://doi.org/10.1007/978-3-030-01941-9_14

located at Bourguiba Avenue. The problem lied in the inadequate choice of the execution technique of piled foundation. In addition, several pathological case histories have been reported for existing buildings founded on Tunis soft clay.

Certainly, the soft soil cannot be extracted without being disturbed. The soft aspect of clay significantly affects the results of laboratory tests mainly because of its sensitivity to transportation. These factors well mark the big difference between results obtained from laboratory tests.

The reconstitution of soft soil represents a suitable solution in using homogeneous samples with controlled water content and well-known history of consolidation. The first, or initial, consolidation allows to avoid the correction of the consolidation curve of reconstituted sample. But in reality, the results obtained from mechanical tests conducted on reconstituted samples are not comparable to results determined from tests performed on intact ground (Bouassida 2006; Klai et al. 2015).

This paper aims to the characterization of Tunis soft clay by compiling collected data from geotechnical surveys conducted for buildings and infrastructures constructed during the last three decades along the Republic avenue of Tunis City. Then correlations are suggested for Tunis soft clay. Using the correlated geotechnical parameters an embankment case study built on Tunis soft clay (TSC) improved by geodrains is investigated numerically. The validation of predicted embankment behavior is discussed.

2 Site and Database Description

The studied area is the Republic Avenue which runs along 2 km. This avenue connects with the highway A1 from the north to the south side. Guilloux and Nakouri (1976) developed geotechnical maps classifying this site as compressible ground. Later, data collected by Kaâniche (1989) classified this site to be with poor characteristics formed by a muddy complex up to 60 m depth. Ammar (1989) developed an electronic geotechnical data describing the lithology as: embankment, soft clay and rigid stratum at 67 m of depth.

Since a decade the study of behavior of Tunis soft clay started at the geotechnical laboratory of National Engineering School of Tunis (Bouassida 1996; Tounekti et al. 2008; Touiti et al. 2009). Those investigations only focused on laboratory tests and the simulation of numerical behavior of Tunis soft clay. Recently, more attention has been given to the characterization of Tunis soft clay for much suitable design of structures built in Tunis City where this problematic soil is frequently encountered. Accordingly, it was thought that the creation of an updated geotechnical database from existing projects at the Republic avenue site will be an efficient tool to the characterization of Tunis soft clay. The geotechnical investigations carried out for seven projects are given in Fig. 1.

Data was collected from sixty (60) Pressuremeter borings (PB), thirty-six (36) Cored borings (CB), seven (7) piezocone and cone penetration tests (PCPT/CPT), six (6) Vane shear tests (VST) and one hundred and fifty-seven (157) Undisturbed Samples (US).

Table 1 summarizes the type, the number of borings and reached depth for each geotechnical survey mainly performed for buildings and infrastructure projects along the Republic Avenue (Tunis City).

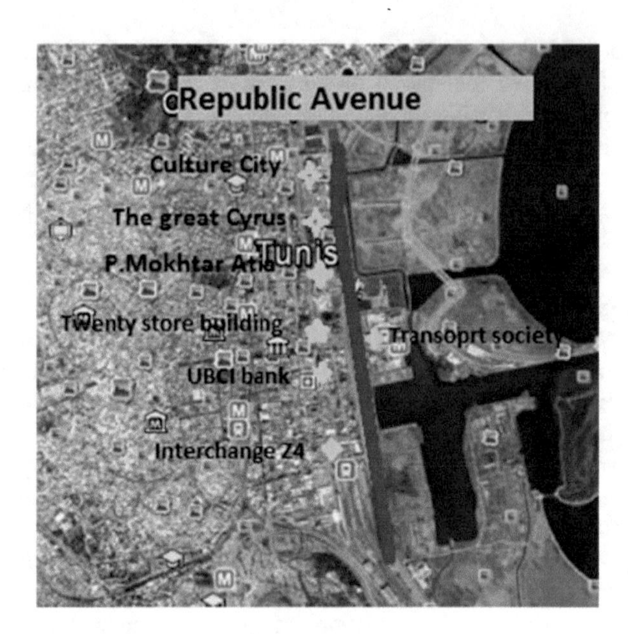

Fig. 1. Map location of the studied area

Table 1. Consistency of in situ geotechnical investigations

Project	Survey		
	Type	Depth (m)	Number
Culture city	PB	61–69	21
	CB	60–70	11
Interchange "Cyrus Le Grand"	PB	65–70	11
	CB	65	9
	PCPT	28	2
Interchange Z4	PB	68–70	15
	CB	70	6
	CPT	40	2
	PCPT	40	3
	VST	15	4
National transportation society	PB	70	3
	CB	70	2
	VST	51–55	2
UBCI bank	PB	60	2
	CB	60	2
Twenty store building	PB	70	6
	CB	25–70	4
Mokhtar ATTIA parking	PB	60	2
	CB	60	2

Statistically, the consistency of site investigation program confirms that the pressuremeter apparatus is the most used testing tool in geotechnical investigation with 55% like proved by Haffoudhi et al. (2005).

The results of laboratory tests were selected over one hundred fifty-seven (157) grain size distributions, twenty-nine (29) oedometer tests, seven (7) direct shear tests, thirteen (13) triaxial CU + u tests and twenty-two (22) triaxial UU tests.

The selected results are compiled to suggest an updated Tunis soft clay characterization with related correlations. Using these results the simulation of behavior of embankment ramps is investigated.

3 Results and Correlations

Results

The averaged soil profile along the Republic avenue shows an upper fill layer of thickness of 2 to 4 m, followed by a soft grayish clay layer of variable thickness from 15 to 20 m having quite low pressuremeter modulus (E_M) between 0.2 and 17 MPa and limit net pressure (Pl^*) between 0.08 and 0.8 MPa, a compressible clay layer with sandy clay to sand in some places having a pressuremeter modulus between 0.5 and 30 MPa and net limit pressure between 0.08 and 1.4 MPa. The rigid stratum is located at 50 m depth toward the North direction while the level stratum is 60 to 70 m in the South direction. The ratio E_M/Pl^* related to the soft clay in the first 20 m depth is generally between 6 and 10. Thus, this soil after Menard's correlation is classified under consolidated. CPT and CPTu data reveal that the soft soil layer is characterized by a tip resistance (q_c) lower than 1 MPa and the friction ratio is between 1 and 3. According to Van Wambeke et al. (1982), the ratio qc/pl is 3 in clays that was confirmed by Frikha et al. (2013).

Referring to the compiled data, the calculated values for the water content are high on the surface with 94%, these values decrease with depth because of the consolidating effect. The liquid limit is usually between 45 and 60%. The plasticity index ranges between 20 and 40%. The specific weight of the solid grains varies from 26 to 27 kN/m^3. These values are closely similar to those initially measured by Bouassida (1996) and, later on, confirmed by Klai (2014) for intact Tunis soft clay specimens.

Correlations

Geotechnical parameters correlations from the database are useful for several objectives; Robertson and Mayne (1998). Firstly, correlations are important for an initial estimate of soil parameters to suggest a preliminary design. Second, they are helpful for evaluating parameters obtained from laboratory test. Moreover, they are helpful for the evaluation of the empirical correlations from in situ tests results.

Pearson's correlation is one of the practical tools for the geotechnical parameters to correlate. Pearson's correlation coefficients of some geotechnical parameters of Tunis soft soil are summarized in Table 2.

This reveals algebraic values which lead to a proportional or inversely proportional agreement between geotechnical parameters. It is obvious that the liquid limit, the water content, the compression index and the void ratio are positively correlated. The

Table 2. Pearson's correlation matrix of geotechnical parameters of Tunis soft clay

	Depth (m)	w	W_l	Ip	γ_s	c_c	Cu vane (kPa)	e_0	C' (kPa)	φ' (°)
Depth (m)										
w	−0,57									
Wl	0,02	0,38								
Ip	0,18	0,08	0,79							
γ_s	−0,13	0,12	0,09	−0,29						
c_c	−0,52	0,78	0,35	0,11	0,38					
Cu vane (kPa)	0,63	−0,32	0,29	0,03	−0,60	−0,31				
e_0	−0,69	0,90	0,19	0,02	−0,15	0,89	−0,38			
C' (kPa)	0,55	−0,77	−0,19	0,24	−0,91	−0,55	1,00	−0,89		
φ' (°)	−0,03	−0,15	−0,02	0,54	0,62	0,30	−1,00	0,67	−0,34	

w: water content; Wl: liquid limit; Ip: plasticity index; γ_s: unit weight of solid particles; c_c: compression index; Cu: undrained cohesion; e_0: void ratio; C': undrained cohesion; φ': drained friction angle.

water content can also be correlated with the undrained cohesion, while the friction angle is positively correlated to the void ratio. The undrained cohesion is significantly negatively correlated with the void ratio and the dry unit weight. Also, the water content, the vane test cohesion and the void ratio are negatively correlated with the depth, reflecting that these parameters decrease proportionally with depth.

Table 3 displays the most important correlations between some geotechnical parameters.

Table 3. New correlations between geotechnical parameters of Tunis soft clay

Parameters	Correlation	R^2
c_c, e_0	$c_c = -0.293\ e_0$	0.801
c_c, γ_d	$c_c = -0.5333\ \gamma_d + 1.009$	0.831
c_c, w	$c_c = 0.008w - 0.038$	0.817
φ', I_p	$\Phi' = -0.031\ I_p^2 + 0.372\ I_p + 18.97$	0.755
C', w	$C' = -28.7\ \mathrm{Ln}(w) + 122.2$	0.601
C', e_0	$C' = 204.8\ e_0^2 - 453.4\ e_0 + 256.1$	0.932
C', c_c	$C' = 731.9\ c_c^2 - 675\ c_c + 157.4$	0.595

4 Study of the Settlement of the Ramp Access of "Cyrus Le Grand" Interchange

The interchange of "Cyrus Le Grand" was built from 2006 to 2008 (Fig. 2). This novel two ways interchange aimed to the increase and the improvement of the fluidity of traffic across Tunis City that was estimated by 120 000 vehicles per day. The full length

Fig. 2. The interchange of "Cyrus Le Grand" in 2007

of the interchange is 475 m, it is composed of two spans of 13 m width each founded on bored piles embedded at 60 m depth.

A monitoring system was implemented to follow-up the evolution of consolidation settlement. Twelve (12) piezometers, 40 rods' settlement, 12 hydraulic settlement and 5 multipoints settlement recorders were installed over the whole area of studied interchange.

In this paper, the main interest is to investigate the evolution of settlement of the ramp access, namely C1, from the North direction to Bizerte City. The soil profile and installed settlement recorders are shown in Figs. 3 and 4.

Geotechnical Profile

The geotechnical profile under the access ramp shows a first fill layer of 4 m thickness followed by soft grayish clay layer of 11 m thickness. Then, a thin sand layer of 4 m thickness followed by a grayish black clay and sandy clay layers of 40 m thickness are crossed up to the top level of rigid stratum layer located at 60 m depth.

Geodrains of 18 m in length were installed in square pattern with axis to axis spacing of 1.1 m to accelerate the consolidation of high compressible soft soil upper layers with compression index of 0.42 (Table 4).

A preloading embankment of 3 m total height was built in two construction stages:

- The first phase refers to the preloading started on May 25th, 2007 and runs for 24 days to reach a height of 2 m.
- The second phase refers to the preloading started on June 29th, 2007 and runs to achieve 95 days to reach a height of 3.1 m.

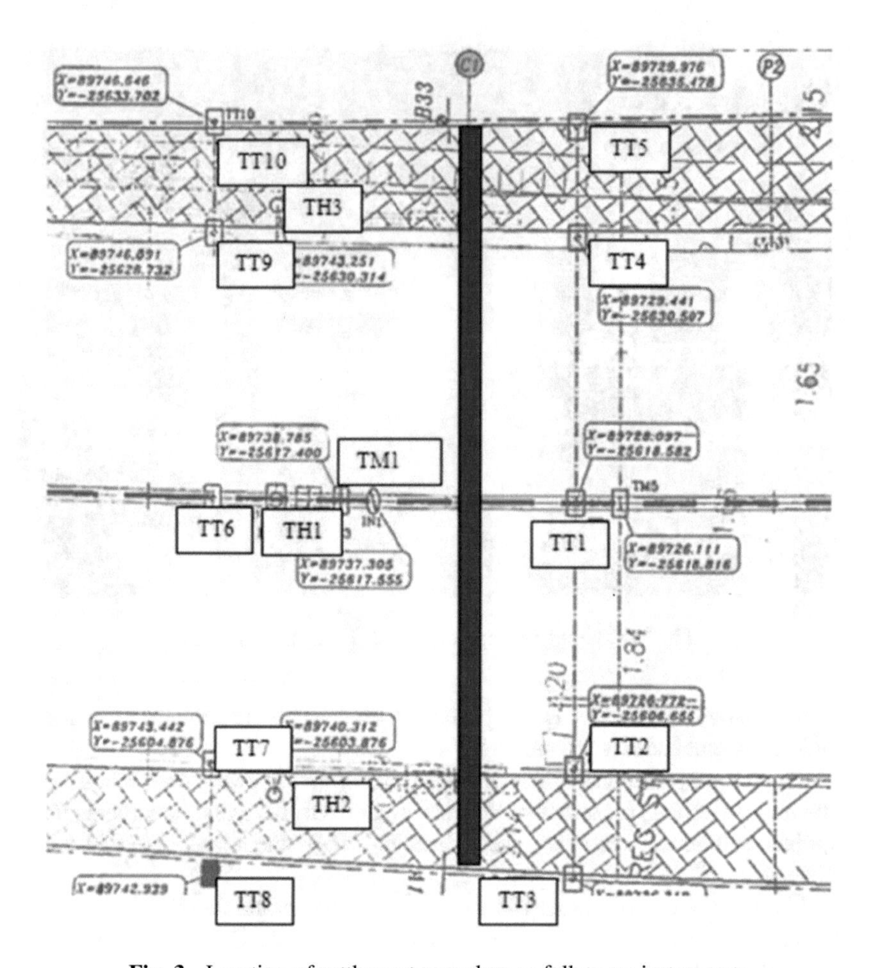

Fig. 3. Location of settlement recorders as follow up instruments

A drainage sand mattress of 0.5 m thickness preceded the embankment construction at the surface of improved soil to facilitate the water evacuation from the geodrains. This sand layer can also contribute of a better load transfer avoiding significant differential settlement.

Evolution of Settlement

The settlement recorders were installed at 5 m in front of the abutment C1 and at 10 m behind it. Rod settlement TT1 was installed at the embankment axis; TT2 and TT3 rods settlement were installed on the right side, whilst TT4 and TT5 were installed on the left side. Behind the abutment C1, rod settlement TT6 was installed at embankment axis, TT7 and TT8 on the right side, TT9 and TT10 on its left side. Hydraulic settlement TH1 and multipoint settlement TM1 were installed at embankment axis, TH2 and TH3 multipoint settlement were installed on the right and on the left side, respectively.

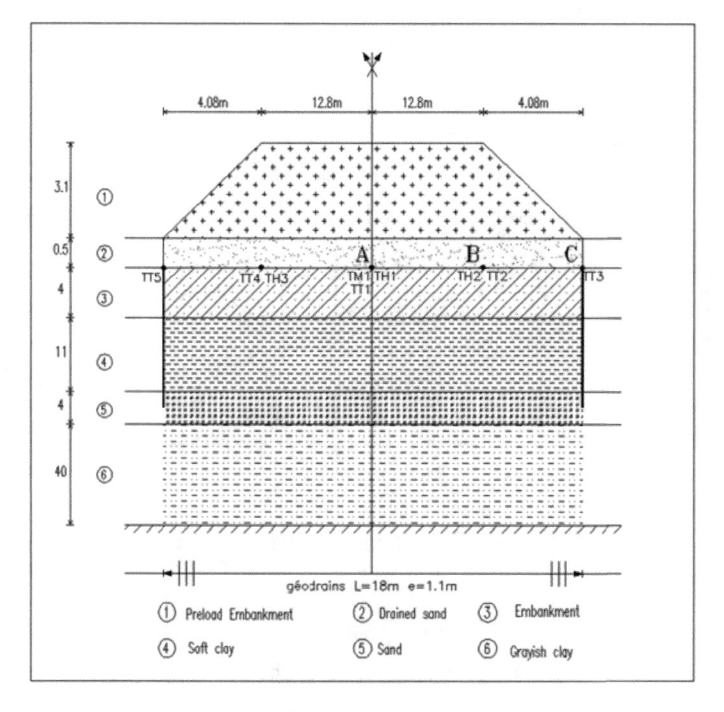

Fig. 4. Geotechnical profile of embankment foundation

Table 4. Correlation between the void ratio and the compression index

Correlation	Author (s)	Year
$c_c = 0.3\ (e_0 - 0.27)$	Terzaghi and Peck	1948
$c_c = 0.54\ (e_0 - 0.35)$	Nishida	1956
$c_c = 0.208\ e_0 + 0.0083$	Azzouz et al.	1976
$c_c = 0.5269\ (e_0 - 0.2117)$	Nath and DeDalal	2004

The follow-up of settlement was maintained for a period of 3 months. Unfortunately, after this short period of time there were no recorded settlements to date. Figures 5, 6, 7 and 8, show the evolution of recorded settlements in the axis and two extremities of instrumented embankment. It can be noted that that the recorded settlement at the axis of embankment were significantly higher than those measured at the embankment extremities. This observation also remains valid for recorded values by the rod settlement and hydraulic settlement tools as illustrated by Figs. 5, 6 and 7, respectively. Due to the symmetrical ramp geometry, the recorded settlements by TT2 and TT4 as well as TT3 and TT5 were almost similar. The significant difference between those recorded settlements can be attributed, first, to the difference of soil profile and, second, for a different spacing of vertical drains that was 1.3 m.

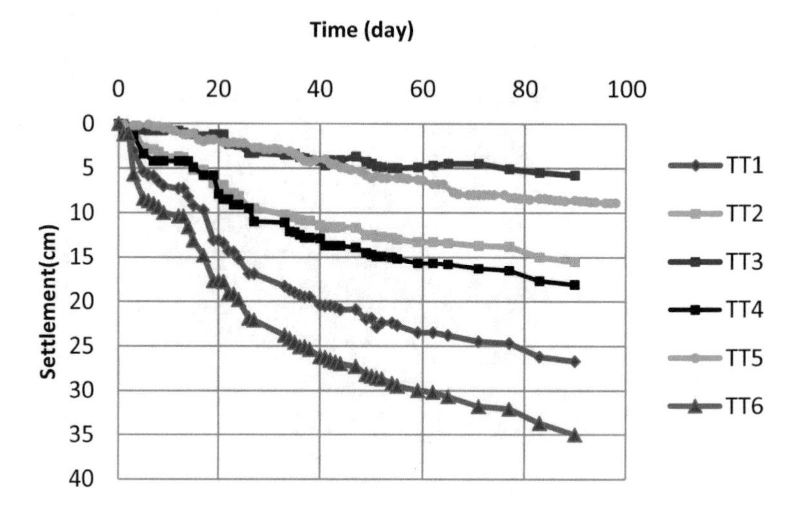

Fig. 5. In situ settlement measurement by rod settlement

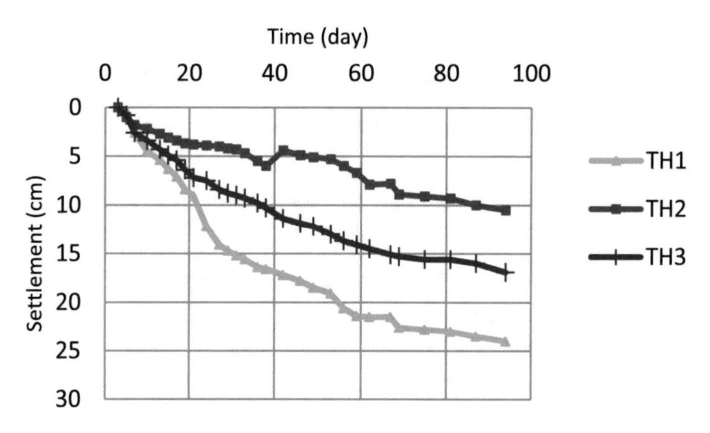

Fig. 6. In situ settlement measurement by hydraulic settlement

Along the embankment axis, it is also noted that the multipoint settlement recorded a value of 37.6 cm after 95 days. However, by the hydraulic settlement the recorded value was limited to 24 cm after 94 days. It was reported by the follow-up team that recorded values provided by the multipoint settlement tool should be handled with caution. Indeed it was observed that the probe of the multipoint settlement tool could not be lowered due to the lateral deformation of the tube (MEHAT 2007).

From recorded values by the rod settlement were comprised between 26.7 cm and 35 cm after 90 days over a distance 15 m between TT1 and TT6 devices. It is concluded that the rod settlement recorded values were in between the hydraulic settlement and the multipoint-settlement measurements.

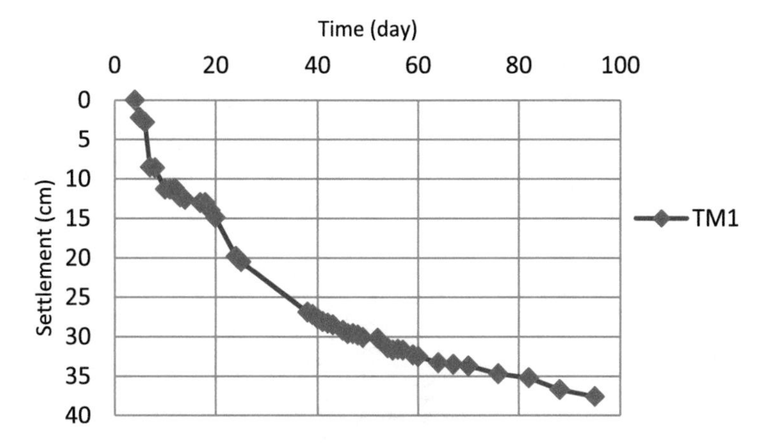

Fig. 7. In situ settlement measurement by multipoint settlement

Numerical Simulation of the Behavior of the Ramp Access

The study of the behavior ramp access embankment was investigated numerically to compare between the recorded vertical displacements and numerical predictions by using the suggested correlation for geotechnical parameters above detailed for Tunis soft clay.

The numerical simulation was run by PLAXIS 2D code (version 2015). A plane strain numerical modeling was adopted. It is composed by 15 nodes triangular finite elements with medium precision mesh to minimize the computation time.

All material characteristics of soil layers and embankment material are given in Table 5. The Mohr Coulomb constitutive law was selected to describe the behavior of embankment material, preload embankment, drained sand and sand layer by adopting a drained behavior. In turn, for the soft soil and grayish-black clay layers, the hardening soil model (HSM) and the soft soil model (SSM) were adopted in undrained behavior.

Related to the boundary conditions of the numerical model, vertical and horizontal displacements are zero at the rigid stratum level; the horizontal displacement along vertical borders is also assumed zero. The water table is located at 2 m depth. The settlement was monitored in accordance with the time given to the actual settlement. The follow up of settlement was investigated in the alignment of three points which are A at the embankment axis, B at the crest of embankment and C at the toe of embankment.

From in-situ observed settlements and predicted ones, it is agreed that the consolidation settlement was increasing versus time, unfortunately only over a period of three months. The in-situ settlement as recorded by all measurement tools fairly show acceptable agreement both in terms of the evolution and magnitude of settlement (Fig. 8).

Referring to the observed settlement, the maximum recorded settlement (after 95 days) by the multipoint settlement recorder at the axis of embankment exceeds the predicted SSM settlement by 5.75 cm which represents 9%. Whereas records by the rod settlement TT1 which represents 15.3% (Fig. 9). Recorded value by the rod settlement TT6 exceeds the predicted SSM settlement by 3.14 cm and the hydraulic

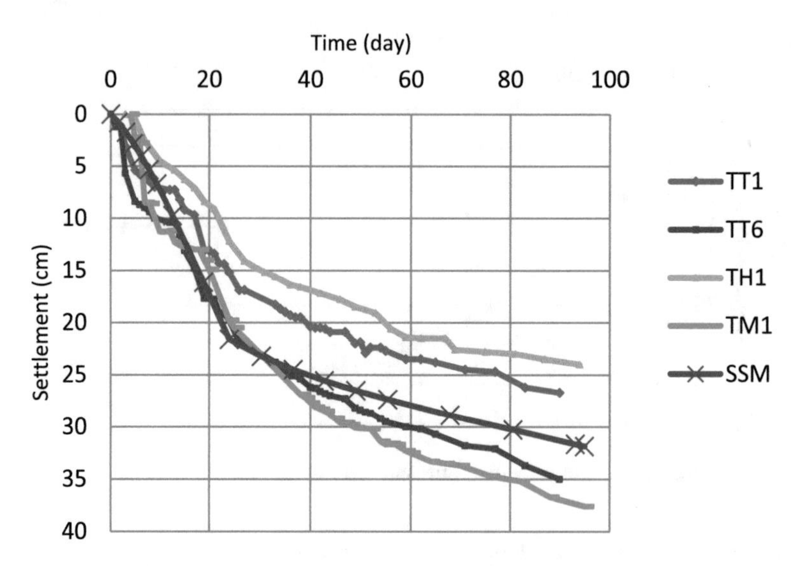

Fig. 8. Evolution of predicted settlement by the SSM and recorded data in the axis

Table 5. Adopted parameters of soft soil and hardening soil models

Parameters	HSM model		SSM model	
	Soft clay	Grayish clay	Soft clay	Grayish clay
γ_h (kN/m^3)	17	18	17	18
γ_{sat} (kN/m^3)	19	20	19	20
$k_x = k_y$ (m/day)	1.52 10^{-4}	1.02 10^{-4}	1.52 10^{-4}	1.02 10^{-4}
E_{50}^{ref} (kN/m^2)	1505.952	1543.421	–	–
E_{oed}^{ref}(kN/m^2)	1204.762	1234.737	–	–
E_{ur}^{ref} (kN/m^2)	8132	7408	–	–
γ_{ur}	0.35	0.38	–	–
R_f	0.9	0.9	–	–
P_{ref} (kPa)	100	100	–	–
c_c	0.42	0.38	0.42	0.38
c_s	0.056	0.057	0.056	0.057
e_0	1.2	1.04	1.2	1.04
C'	–	–	6.08	8.46
φ'(°)	–	–	20.41	20.56
ψ(°)	–	–	0	0

c_s: swelling index; ψ (°): angle of dilatancy; $k_x(_y)$: horizontal (vertical) permeability.

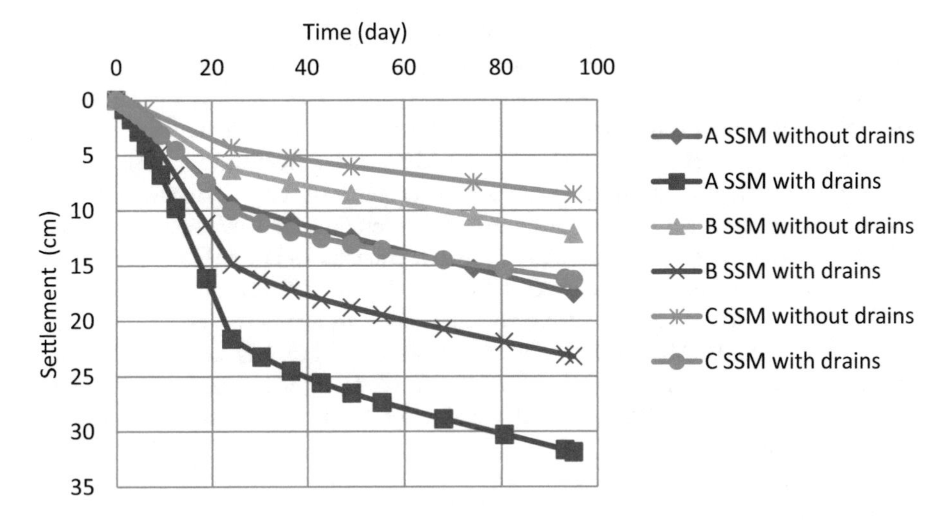

Fig. 9. Numerical settlement predicted by the SSM for improved soil by geodrains and unimproved soil

settlement gave lower values than to the predicted SSM settlement by 5.2 cm and 7.86 cm, respectively. It can be concluded that numerical predictions by the SSM were in a good agreement with the in-situ measurement values which prove the efficiency of correlated parameters.

The behavior of unimproved soil (without geodrains) was also analyzed by the same numerical model to highlight the benefits of geodrains technique (Fig. 9).

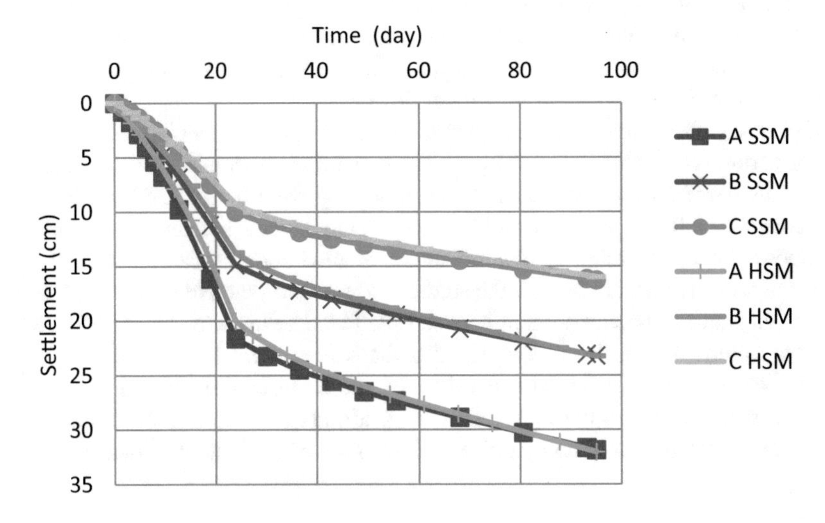

Fig. 10. Numerical settlement obtained by SSM and HSM for improved soft soil by geodrains

The evolution of settlement at embankment axis (point A) predicts 9 cm after 13 days for improved soil by geodrains. But, the same settlement is expected after 24 days for the unimproved soil. The final predicted settlement after 95 days by the SSM is 31.9 cm for the improved soil whilst it is 17.7 cm for the unimproved case. After 95 days, the settlement of unimproved soil is by 55% of that predicted for improved soil by geodrains installed in square pattern with spacing of 1.1 m. Hence, this study confirmed the beneficial role of geodrains in accelerating the consolidation of Tunis soft clay settlement as reported recently from experimental work by (Jebali et al. 2017).

The difference between predicted settlements by the soft soil and hardening soil models does not exceed 2 cm. Such result indicates that by the two constitutive laws, governed by the HSM and SSM, quasi identical behavior is expected to occur (Fig. 10). Consequently, it can be concluded that modeling the behavior of Tunis soft clay by the SSM and the HSM are both suitable for predicting the studied behavior of ramp access embankment.

5 Conclusions

This paper dealt with the characterization of Tunis soft clay and the study of behavior of improved soft soil by geodrains. After geotechnical investigations conducted for seven projects at the site of Republic Avenue of Tunis City selected results were compiled to characterize the geotechnical parameters of Tunis soft soil. Those data led to the suggestion of some correlations between geotechnical parameters of Tunis soft clay. These correlations remain valid for the studied area, therefore their use must be handled carefully for similar projects even around the investigated.

The settlement of ramp access embankment built on improved soft layers by geodrains of 18 m length was monitored by different types of settlement recorders, e.g. rod settlement, hydraulic settlement and multipoint settlement. It was noted that recorded settlement values by the multipoint device slightly exceeded the settlements recorded the rod and hydraulic settlement devices.

Using a plane strain model, a numerical simulation was performed for the validation of observed behavior of ramp access embankment in terms of settlement evolution. The hardening soil and the soft soil modeling were both considered to describe the behavior of Tunis soft clay and compressible layers. Numerical predictions of settlements under the ramp access embankment in different locations revealed quasi similar. It was then concluded, for the studied case history, the HSM and SSM revealed both valid. Indeed, the comparison between predicted settlements by the SSM and the recorded values during the follow-up of embankment were in acceptable agreement especially when recorded settlement values by the multipoint device were considered. It was also checked that the installation of geodrains provide a good acceleration of consolidation settlement in comparison to the simulated settlement of unimproved soil.

So, the correlated geotechnical parameters in the area of Avenue Republic of Tunis City revealed quite helpful to characterize the soft soil layers extending at 40 m depth. This finding was evidenced by the validation of numerical predictions of the behavior of access ramp embankment built of improved soft layers by geodrains.

References

Ammar, B.: Etude de l'environnement de Tunis et de sa région en relation avec l'art de l'ingénieur en génie civil. Thèse de doctorat. Université des Sciences et Technologies, Languedoc, France (1989), 296 p

Azzouz, A.S., Krizek, R.J., Corotis, R.B.: Regression analysis of soil compressibility. Soils Found. **16**(2), 19–29 (1976)

Bouassida, M.: Etude expérimentale de la vase de Tunis par colonnes de sable- Application pour la validation de la résistance en compression théorique d'une cellule composite confinée. Revue française de géotechnique **75**, 3–12 (1996)

Bouassida, M.: Modeling the behavior of soft clays and new contributions for soil improvement solutions. In: Bujang, Pinto, M., Jefferson, I. (eds.) Keynote Lecture: 2nd International Conference on Problematic Soils. Petaming Jaya, Salengor, Malaysia, pp. 1–12 (2006)

Frikha, W., Ben Salem, Z., Boussida, M.: Estimation of Tunis soft soil undrained shear strength from pressumeter data. In: Proceeding of the 18th International Conference on Soil Mechanics and Geotechnical Engineering. Paris (2013)

Guilloux, A., Nakouri, L.: Contribution à l'étude géotechnique des sous-sols de Tunis. Revue Tunisienne de l'Equipement, Tunisie **16**, 58–68 (1976)

Haffoudhi, S., Khediri, S., Zaghouani, K.: Utilisation du pressiomètre en Tunisie. Rapports nationaux. In: Proceedings of ISP5- PRESSIO 2005. International Symposium 50 years of Pressure Meters, 22–24 August 2005, vol. 2, pp. 283–292. Edit. LCPC-ENPC, Paris (2005)

Jebali, H., Frikha, W., Bouassida, M.: 3D consolidation of Tunis soft clay improved by geodrains. Geotech. Test. J. **40**(3), 361–370 (2017)

Kaâniche, A.: Conception et réalisation d'une base de données géologiques (Tunis-Data-Bank) orienté vers la cartographie géotechnique automatique (Tunis, Géo-Map); Application à la ville de Tunis, Thèse de doctorat, Institut National des Sciences Appliquées, Lyon, France, 327 p (1989)

Klai, M., Bouassida, M., Tabchouche, S.: Numerical modelling of Tunis soft clay. Geotech. Eng. J. SEAGS & AGSSEA **46**(4), 87–95 (2015)

Klai, M., Bouassida, M.: Study of the behavior of Tunis soft clay. Innov. Infrastruct. Solut. **1**, 31 (2016). https://doi.org/10.1007/s41062-016-0031-x, http://rdcu.be/ntus

MEHAT: Travaux de l'échangeur Cyrus Le grand-Ghana Liaison Nord-Sud. Rapport de suivi de consolidation des rampes. Direction générale des Ponts et Chaussées, Tunis (2007)

Nath, A., DeDalal, S.S.: The role of plasticity index in predicting compression behavior of clays. Electron. J.Geotech. Eng., 1–7 (2004)

Nishida, Y.: A brief note on the compression index of soil. J. Soil Mech. Found. Div. ASCE **82** (3), 1–14 (1956)

Robertson, P.K., Mayne, P.W.: Geotechnical site characterization. In: Proceedings of the First International Conference on Site Characterization ISC 98. Atlanta, Georgia, 19–22 Apr 1998

Terzaghi, K., Peck, R.B.: Soil Mechanics in Engineering Practice. Wiley, New York (1948)

Tounekti, F., Bouassida, M., Klai, M.: Etude expérimentale en vue d'un modèle de comportement pour la vase de Tunis. Rev. Fr. Géotech. **114**(122), 25–36 (2008)
Touiti, L., Bouassida, M., Van Impe, W.: Discussion on Tunis soft soil sensitivity. Geotech. Geol. Eng. **27**, 631–643 (2009). https://doi.org/10.1007/s10706-009-9263-2

Drained Response of Granular Material

Hoang Bao Khoi Nguyen[1(✉)] and Mizanur Rahman[2]

[1] School of Natural and Built Environments, University of South Australia, Adelaide, Australia
Khoi.Nguyen@unisa.edu.au
[2] Geotechnical Engineering, School of Natural and Built Environments, University of South Australia, Adelaide, Australia
Mizanur.Rahman@unisa.edu.au

Abstract. Drained triaxial simulations were carried out for three-dimensional (3D) assembly of ellipsoid particles using the discrete element method (DEM). It was found that the overall drained behaviour is dependent on both density state i.e. void ratio (e) and stress state i.e. confining stress (p') as observed in many laboratory triaxial tests. The characteristic (Ch) state, where the behaviour (volumetric strain, ε_v) changes from contractive to dilative, was also observed for dilatant specimen. All these simulations reached critical state (CS) at the end of their simulation. The CS data points formed a unique CS line (CSL) in e-log (p') and q-p' space regardless of the initial states. The CSL facilitated further analysis in term of the state parameter (ψ), which is the difference in e and e at CS for the same p'. It was found that Ch states form a unique relationship with the ψ. This enabled to adopting a state dependent flow rule to capture dilatancy behaviour. This allowed adopting a SANISAND model to predict drained behaviour. Despite some limitations, the model was able to predict overall trends of shearing response. This study enhances the understanding of fundamental basis of the relationship between ψ and mechanical parameters of a cohesionless soil.

1 Introduction

The drained response of granular materials has been commonly classified as contractant (C) or dilatant (D) behaviour as shown in Fig. 1. A loose specimen often exhibits C behaviour, whereas D behaviour is commonly observed in a dense specimen. When shearing loose specimen, volumetric contraction with deviatoric stress (q) hardening occurs throughout this stage until reaching an equilibrium state i.e. $d\varepsilon_v = dq = 0$, where ε_v is volumetric strain, $q = (\sigma'_1 - \sigma'_3)$, σ'_1 and σ'_3 are major and minor effective principal stresses respectively. For D behaviour, q hardening takes place until reaching initial peak strength, and then q softens toward an equilibrium. In this case, the dense specimen initially exhibits volumetric contraction and then phase-transforms to volumetric dilation. The transition between volumetric contraction and dilation is referred to as characteristic (Ch) state. Ch state, sometime refer to as phase transformation state, is considered as a key characteristic that control dilatancy of granular materials, and has been used in bounding surface modelling. An equilibrium state of effective stress or volumetric strain at later stage of shearing is often called critical state (CS), which is an

S. Hemeda and M. Bouassida (Eds.): GeoMEast 2018, SUCI, pp. 175–184, 2019.
https://doi.org/10.1007/978-3-030-01941-9_15

anchor concept of critical state soil mechanics (CSSM) framework. The CS line (CSL), as defined by a set of CS data points in e-log(p') space, is also a reference state for modelling C and D behaviour e.g. C behaviour is observed for a soil state above the CSL, and D behaviour and Ch state are observed for a state below the CSL; where e is void ratio and p' is mean effective stress. The state parameter (ψ), which is the difference between e at current state and e on the CSL at the same p' (Been and Jefferies 1985), is a mathematical expression of such a state index and can be easily implemented in a constitutive model formulation.

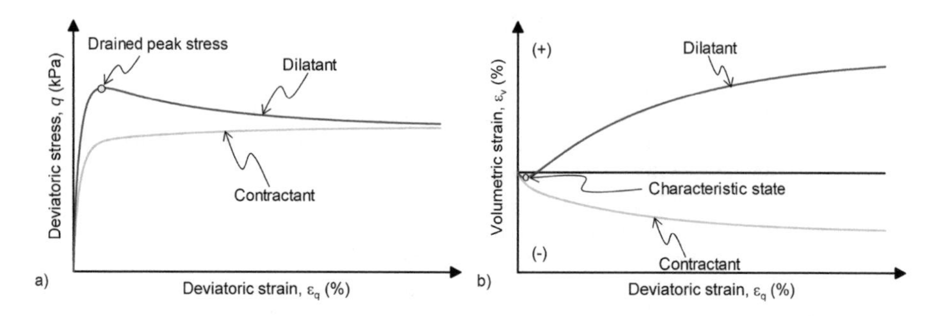

Fig. 1. Schematic diagram of observed drained response of granular material.

Discrete element method (DEM) has recently become a common approach for qualitative understanding of soil behaviour under a CSSM framework (Huang et al. 2014; Nguyen and Rahman 2017; Nguyen et al. 2015a, b), as it does not suffer from experimental limitation of non-uniformity and allows the understanding of internal structure of soil. There have been previous studies on CS behaviour of the assembly of idealized particles e.g. discs or balls (Huang et al. 2014; Sitharam and Vinod 2009; Zhao and Guo 2013); however, natural soils often have some degree of angularity. Therefore, this study adopted the assemblies of 3D ellipsoid particles to observe the response of granular material under drained condition and also capture some angularity features of real soils. The Ch state and CS were also evaluated to understand their constitutive relations with ψ. Then, a state-dependent constitutive model by Li and Dafalias (2000) was calibrated to understand the challenges in DEM simulation with the existing elasto-plastic formulations of continuum mechanics. This study thus enhances the understanding of the fundamental basis of the relationship between ψ and mechanical response of cohesionless granular materials.

2 Dem Simulation for Drained Condition

2.1 Preparation Method

The details of the specimen for this study can be also found in Nguyen et al. (2018). The numerical control for drained simulations is exactly same as actual tri-axial testing condition. Hence, there is no pore water pressure development. A constant incremental rate for axial strain (ε_1) is applied, while the rate of change of σ'_3 is kept constant i.e. $d\sigma'_3 = 0$ to achieve the drained path in effective stress space; where notation 'd' indicates the rate of change. Note, the drained path in the effective stress space has a ratio of $dq/dp' = 3$.

2.2 Observed Drained Behaviour

Figure 2 shows the observed responses to drained shearing of one dense (DISO4, $e_0 = 0.582$ & $p'_0 = 50$ kPa) and one loose (DISO14, $e_0 = 0.670$ & $p'_0 = 100$ kPa) specimens. It should be noted that deviatoric strain ($\varepsilon_q = 2/3[\varepsilon_1 - \varepsilon_3]$) and volumetric strain ($\varepsilon_v = \varepsilon_1 + 2\varepsilon_3$) are considered for this investigation; where ε_1 and ε_3 are axial and lateral strains respectively. The dense specimen exhibits D behaviour, where q hardening initially takes place and then q softening occurs after reaching an initial peak. According to Li and Dafalias (2000) at the initial peak of q, the plastic hardening modulus (K_p) becomes 0, which is important to calibrate the SANISAND model parameters. The loose specimen exhibits C behaviour which associates with q hardening throughout shearing.

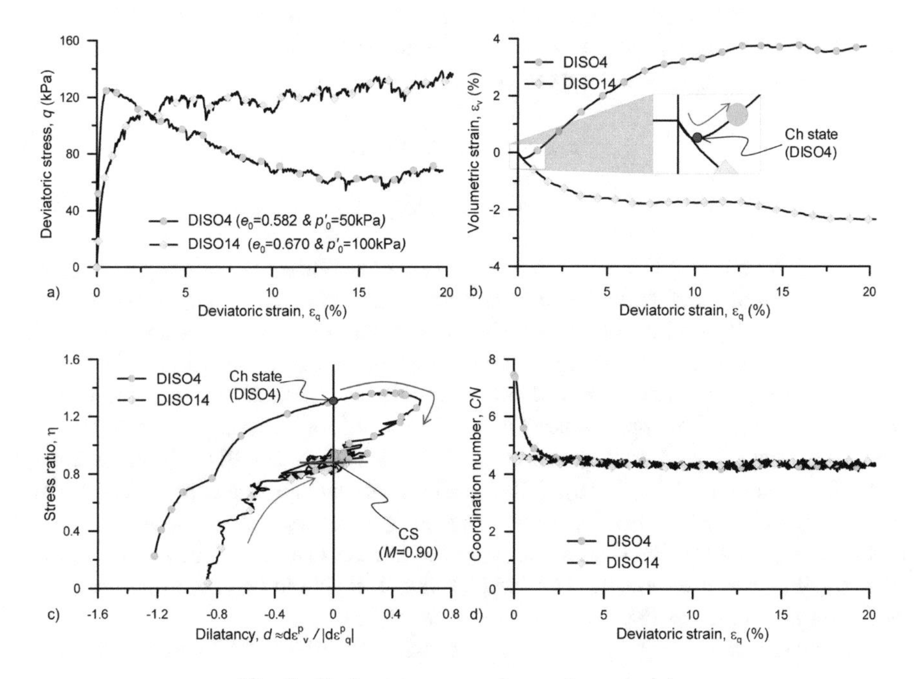

Fig. 2. Drained response of granular material.

In Fig. 2b, DISO14 shows volumetric contraction, whereas DISO4 initially exhibits volumetric contraction then shifts phase to volumetric dilation; note (−) sign means contraction and (+) sign mean dilation in this study. The Ch state of DISO4, where local minimum of ε_v happens, occurs in the early stage of shearing. At this point, $d\varepsilon_v$

also becomes 0. The dilatancy behaviour of these two simulations are also shown in Fig. 2c in terms of stress ratio (η) – dilatancy (*d*) relation (Been and Jefferies 2004; Li and Dafalias 2000); where *d* is the ratio between the rate of plastic ε_v to plastic ε_q i.e. $d = d\varepsilon_v^p / \left|d\varepsilon_q^p\right|$. According to Fig. 2c, the dilatancy path of DISO4 gradually evolves to a positive *d* from a negative *d* and then comes back to 0 (at CS), whereas the path of DISO14 stops increasing at $d = 0$ (at CS). The first time, when the dilatancy path of a dense specimen (DISO4) reaches $d = 0$, is at Ch state. Then, the second time this happens is at CS. The critical state stress ratio ($M = q_{CS}/p'_{CS}$) for all simulations in this study is 0.90.

Figure 2d shows the evolution path of coordination number (*CN*) during shearing. DISO4 initially has higher *CN* than DISO14. However, as ε_q increases, *CN* values of DISO4 quickly drops and becomes identical to that of DISO14.

2.3 Critical State Behaviour

As discussed previously, CS is achieved at which there is no change in volume (or *e*) and stresses with a continuous shearing. According to the CSSM framework, a set of CS data points can form a unique CSL in the classical *e*-log(p') space and the critical state stress ratio (*M*) in the *q*-p' space (Schofield and Wroth 1968). The CSL in the classical *e*-log(p') space can be expressed as:

$$e_{CS} = e_{lim} - \Lambda (p'/p_a)^{\xi} \tag{1a}$$

$$M = q_{CS}/p'_{CS} \tag{1b}$$

where e_{lim}, Λ and ξ are CS parameters which are controlling the y-intercept, slope and curvature of CSL; and *M* is the critical state stress ratio. p_a is atmospheric pressure which is 101 kPa. A series of drained simulations are done to define a unique CSL for this granular material. The details of these simulations can be also found in Nguyen et al. (2018). Some simulation did not reach clear CS; therefore, the extrapolation method was adopted to approximate the CS. The details of this method can be found in some previous CS studies (Nguyen et al. 2017; Rabbi et al. 2018; Rahman and Lo 2014; Rahman et al. 2017, 2018). CSL for all drained simulations is plotted in a three-dimensional *q*-*e*-log(p') space as shown in Fig. 3; where e_{lim}, Λ and ξ are 0.68, 0.02 and 0.73 respectively. The CSL is the reference line for C and D responses of granular material, which is used for the SANISAND model. It should be noted that the curved *M*-line in *q*-p' space in Fig. 3 is due to the log scale of p'.

The state parameter (ψ) proposed by Been and Jefferies (1985) can be defined as:

$$\psi = e_{CS} - e \tag{2}$$

In the literatures, ψ has been commonly used as the state index for granular material. The relationships between ψ and other parameters related to material behaviour (e.g. instability stress ratio, cyclic resistance ratio, etc.) have also been widely

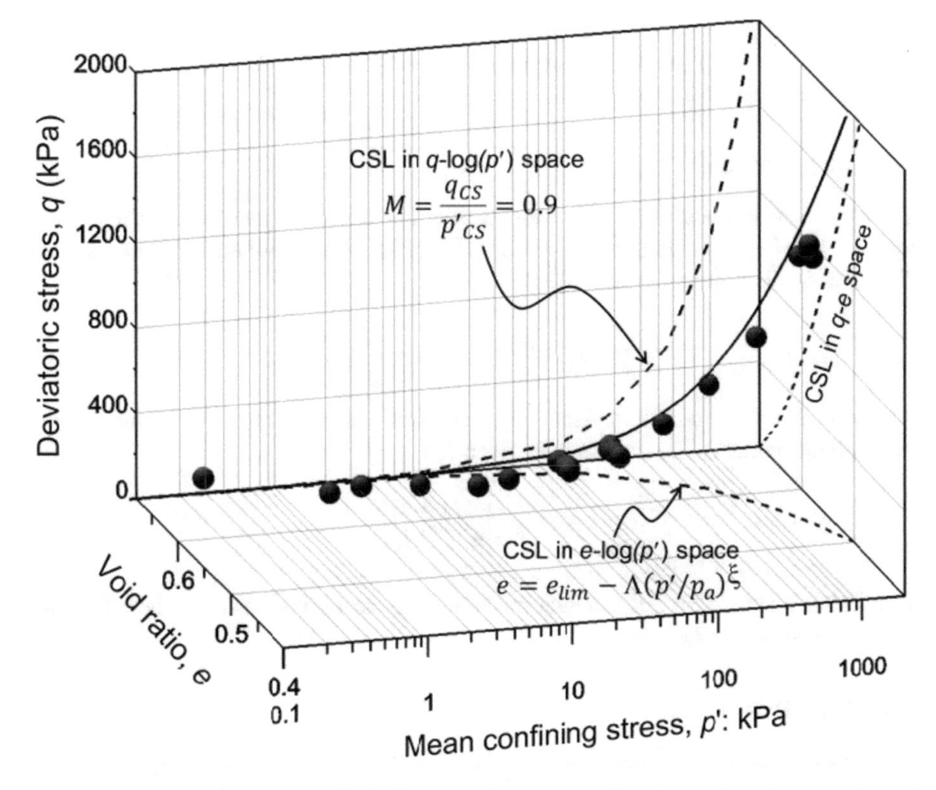

Fig. 3. CSL in *q*-*e*-log(*p*′) space.

investigated. A state-dependent constitutive model based on ψ, SANISAND (Li and Dafalias 2000), is adopted to calibrate the DEM simulations in this study. Accordingly, the challenges in DEM simulation with the existing elasto-plastic formulations of continuum mechanics will be explored.

3 Constitutive Formulation

For a simple triaxial model, Li and Dafalias (2000) approximated the yield function as a family of constant stress ratio lines, hence:

$$f(q, p = q - \eta\, p' = 0) \tag{3}$$

The additive decomposition of strain increment, as in one of SANISAND model (Li and Dafalias 2000), is often considered as:

$$\mathrm{d}\varepsilon_{ij} = \mathrm{d}\varepsilon_{ij}^{e} + \mathrm{d}\varepsilon_{ij}^{p} \tag{4}$$

where superscripts "e" and "p" denote elastic and plastic components, respectively. The model consists of a series of constitutive equations for elastic and plastic strain increment components.

3.1 Elastic Moduli

The elastic shear modulus (G) of a sandy soil can be expressed as a function of e and p' by the following equation (Goudarzy et al. 2017; Hardin and Richart 1963; Iwasaki and Tatsuoka 1977; Rahman et al. 2012):

$$G = G_0 \left\lfloor (2.97 - e)^2/(1 + e) \right\rfloor \sqrt{p' p_a} \tag{5}$$

where G_0 is model parameter. Assuming a constant Poisson's ratio (ν) the incremental elastic bulk modulus (K) for sands is given by:

$$K = G \frac{2(1 + \nu)}{3(1 - 2\nu)} \tag{6}$$

ν of 0.05 is assumed for this granular material, whereas G_0 is determined by calibrating the q-ε_q at small strain

3.2 Flow Rule

The flow rule in this model is a modification of Rowe's stress dilatancy equation based on considerations of CSSM via its dependence on ψ. More specifically the dilatancy equation for the clean sand in Li and Dafalias (2000, 2012) is given by:

$$d = \frac{d\varepsilon_v^p}{|d\varepsilon_q^p|} = d_0 \left[M e^{m\psi} - \eta \right] / M \tag{7}$$

where d_0 and m are model parameters assumed to be independent of e and ψ. As mentioned previously, $d = 0$ at the Ch state. By re-arranging Eq. (7), the m can be obtained by this relation:

$$\eta_{Ch} = M e^{m\psi_{Ch}} \tag{8}$$

where subscript 'Ch' indicates the Ch state. The relation in Eq. (8) is also presented in Fig. 4; which was very similar to recently observed experimental data by Zhang et al. (2018). The Eq. (8) gave an absolute value of m of 2.65. After defining m, d_0 can be acquired by calibrating the dilatancy curve (η-d relation).

3.3 Bounding Surface and Plastic Modulus

Li and Dafalias (2000) suggested the incremental stress ratio-plastic deviatoric strain relation which can be written as:

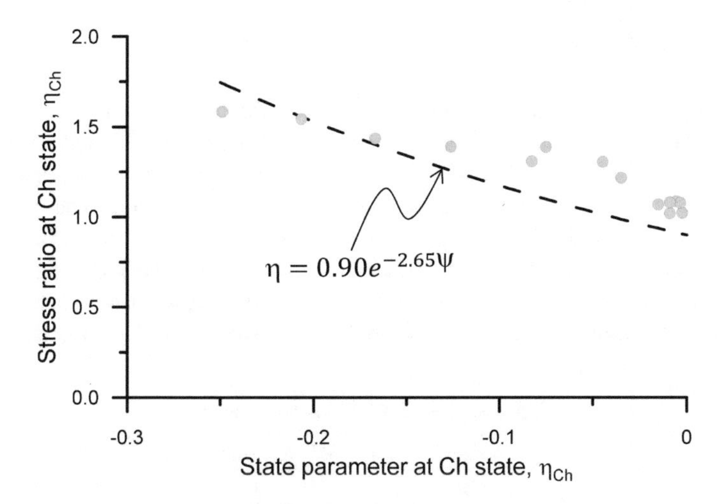

Fig. 4. Relationship between η and ψ at the Ch state.

$$d\eta/d\varepsilon_q^p = K_p/p' \tag{9}$$

where K_p is the plastic hardening modulus given by:

$$K_p = hG\left(M/\eta - e^{n\psi}\right) \tag{10}$$

where n and h are positive model parameters. h can be expressed as:

$$h = h_1 - h_2 e \tag{11}$$

where h_1 and h_2 model parameters assumed to be independent of density. As mentioned previously, K_p at initial q peak in the q-ε_q curve of D behaviour becomes 0. Therefore, the Eq. (10) can be re-arranged and n can be defined as following:

$$\eta_p = M/e^{n\psi_p} \tag{12}$$

where subscript 'p' indicates the state at which the initial peak q occurs. After obtaining n, the other two hardening parameters (h_1 and h_2) can be determined by calibrating the q-ε_q curve.

3.4 Increment Equations

The strain increment equations can be expressed as:

$$d\varepsilon_q = d\varepsilon_q^e + d\varepsilon_q^p = dq/G + p'd\eta/K_p \tag{13}$$

$$d\varepsilon_v = d\varepsilon_v^e + d\varepsilon_v^p = dp'/K + dd\varepsilon_q^p \tag{14}$$

4 Model Calibration

The model parameters obtained from previous discussions are listed in Table 1. These parameters can be also found in Nguyen et al. (2018). The prediction model for DISO4 and DISO14 are MDISO4 and MDISO14 respectively, which are shown in Fig. 5. The q-ε_q relations of the two simulations are well predicted by this model. The ε_v-ε_q relation of DISO4 also fit well with the prediction model. However, the prediction model MDISO14 is quite different to DISO14. This discrepancy may be due to the calibration of d_0. Despite the issue when fitting the ε_v-ε_q path, the simple elasto-plastic model in continuum mechanics shows the capability of predicting a DEM simulation.

Table 1. SANISAND model parameters

Parameter	Symbol	Value
Elastic	ν	0.05
	G_o (MPa)	75
Critical state	M	0.9
	e_{lim}	0.68
	Λ	0.02
	ξ	0.73
Dilatancy	d_0	0.80
	m	2.65
Hardening	h_1	3
	h_2	2
	n	4.5

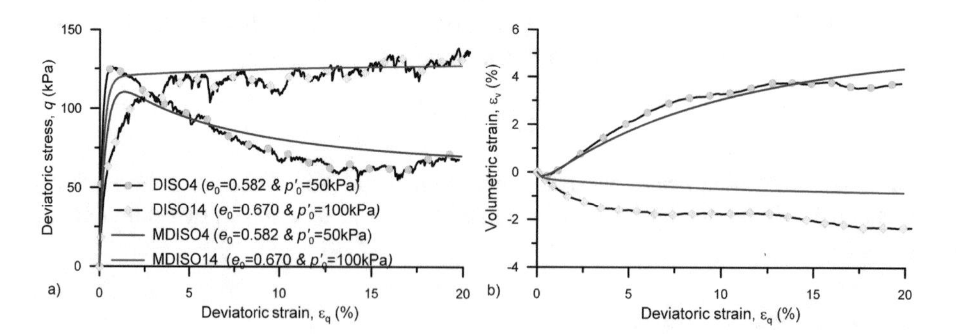

Fig. 5. Prediction model for the drained simulations.

5 Conclusions

The major findings of this study are discussed as the followings:

- The drained response of granular material in this study shows good agreement with the literatures, which proves that DEM can be an effective tool to capture the qualitative behaviour of granular material.
- A unique CSL is obtained from a series of drained simulations. This reference line is adopted as state indicator (contractant or dilatant behaviour) in SANISAND model.
- The hardening parameter (n) is calculated in the drained peak stress. Other hardening parameters are calibrated in the q-ε_q plot. These parameters help to effectively capture the q hardening and softening paths of DEM simulations.
- The dilatancy parameter (m) is determined at Ch state, whereas another dilatancy parameter (d_0) is calibrated in the ε_v-ε_q plot. But the model cannot predict well the ε_v of C behaviour. It may be due to the consideration of m as discussed previously.

References

Been, K., Jefferies, M.: Stress dilatancy in very loose sand. Can. Geotech. J. **41**(5), 972–989 (2004)

Been, K., Jefferies, M.G.: A state parameter for sands. Geotechnique **35**(2), 99–112 (1985)

Goudarzy, M., Rahemi, N., Rahman, M.M., Schanz, T.: Predicting the maximum shear modulus of sands containing nonplastic fines. J. Geotech. Geoenviron. Eng. **143**(9), 06017013 (2017). https://doi.org/10.1061/(ASCE)GT.1943-5606.0001760

Hardin, B.O., Richart, F.E.J.: Elastic wave velocities in granular soils. J. Soil Mech. Found. Div. **89**(1), 33–65 (1963)

Huang, X., O'Sullivan, C., Hanley, K., Kwok, C.: Discrete-element method analysis of the state parameter. Geotechnique **64**(12), 954–965 (2014)

Iwasaki, T., Tatsuoka, F.: Effects of grain size and grading on dynamic shear moduli of sands. Soils Found. **17**(3), 19–35 (1977)

Li, X., Dafalias, Y.: Dilatancy for cohesionless soils. Geotechnique **50**(4), 449–460 (2000)

Li, X., Dafalias, Y.: Anisotropic critical state theory: role of fabric. J. Eng. Mech. **138**(3), 263–275 (2012). https://doi.org/10.1061/(ASCE)EM.1943-7889.0000324

Nguyen, H.B.K., Rahman, M.M.: The role of micro-mechanics on the consolidation history of granular materials. Aust. Geomech. **52**(3), 27–36 (2017)

Nguyen, H.B.K., Rahman, M.M., Cameron, D.A.: Undrained behavior of sand by DEM study. In: International Foundations Congress Equipment Expo (IFCEE 2015), pp. 182–191. ASCE (2015a). https://doi.org/10.1061/9780784479087.019

Nguyen, H.B.K., Rahman, M.M., Cameron, D.A., Fourie, A.B.: The effect of consolidation path on undrained behaviour of sand - a DEM approach. Computer Methods and Recent Advances in Geomechanics, CRC Press, pp. 175–180 (2015b). https://doi.org/10.1201/b17435-27

Nguyen, H.B.K., Rahman, M.M., Fourie, A.B.: Undrained behaviour of granular material and the role of fabric in isotropic and K_0 consolidations: DEM approach. Géotechnique **67**(2), 153–167 (2017). https://doi.org/10.1680/jgeot.15.P.234

Nguyen, H.B.K., Rahman, M.M., Fourie, A.B.: Characteristic behaviour of drained and undrained triaxial tests: a DEM study. J. Geotech. Geoenviron. Eng. (2018, in Press). https://doi.org/10.1061/(asce)gt.1943-5606.0001940

Rabbi, A.T.M.Z., Rahman, M.M., Cameron, D.A.: Undrained behavior of silty sand and the role of isotropic and K_0 consolidation. J. Geotech. Geoenviron. Eng. **144**(4), 1–11 (2018). https://doi.org/10.1061/(ASCE)GT.1943-5606.0001859

Rahman, M., Lo, S.: undrained behavior of sand-fines mixtures and their state parameter. J. Geotech. Geoenviron. Eng. **140**(7), 04014036 (2014). https://doi.org/10.1061/(ASCE)GT.1943-5606.0001115

Rahman, M.M., Cubrinovski, M., Lo, S.R.: Initial shear modulus of sandy soils and equivalent granular void ratio. Geomech. Geoengin. **7**(3), 219–226 (2012). https://doi.org/10.1080/17486025.2011.616935

Rahman, M.M., Nguyen, H., Rabbi, Z.: Undrained behaviour of sand under isotropic and K_0-consolidated condition: experimental and DEM approach. In: The 19th International Conference on Soil Mechanics and Geotechnical Engineering, pp. 493–496 (2017)

Rahman, M.M., Nguyen, H.B.K., Rabbi, A.T.M.Z.: The effect of consolidation on undrained behaviour of granular materials: a comparative study between experiment and DEM simulation. Geotech. Res. (2018, in press)

Schofield, A.N., Wroth, P.: Critical State Soil Mechanics. McGraw-Hill, London (1968)

Sitharam, T., Vinod, J.S.: Critical state behaviour of granular materials from isotropic and rebounded paths: DEM simulations. Granul. Matter **11**(1), 33–42 (2009)

Zhang, J., Lo, S.-C.R., Rahman, M.M., Yan, J.: Characterizing Monotonic behavior of pond ash within critical state approach. J. Geotech. Geoenviron. Eng. **144**(1), 04017100 (2018). https://doi.org/10.1061/(ASCE)GT.1943-5606.0001798

Zhao, J., Guo, N.: Unique critical state characteristics in granular media considering fabric anisotropy. Géotechnique **63**(8), 695–704 (2013)

Cut Slope Stability Analysis of Rangvamual Landslide Along Aizawl Airport Road, Northeast India

Lal Dinpuia(✉)

Geology Department, Pachhunga University College (Mizoram University), Aizawl, India
laldinpuia@pucollege.edu.in

Abstract. Rangvamual landslide covers an area of 15,316.4 m^2, at the western limb of Aizawl anticline along NH-54, the State of Mizoram lifeline and airport road. Landslide experienced in the area after slope modification for building construction followed by rainfalls on June and September in 2014. One building collapsed, landslide debris affected Cold Storage Building, distorted one post of 132 kV power line that connecting western towns of the State, and frequently interrupted traffic for more than 2 years. The landslide can be classed as 'debris slide and earth slide' of translational type. Proper mitigation measures are suggested after geotechnical and Limit Equilibrium Method (LEM) analyses.

1 Introduction

The transportation systems and construction activities in the hilly regions are implemented without proper understanding of geological and geotechnical conditions of the slopes is one of the major cause of the failures encountered in these areas (Singh et al. 2013, 2016). Limit Equilibrium Method (LEM) is widely used by researchers and engineers for slope stability analysis (Cheng et al. 2007).

Rangvamual landslide situated at the western limb of Aizawl anticline along NH-54. The failure occurred after unscientifically slope cut activities, and triggered by incessant rainfalls (1371.1 mm during May to September; Source: Meteorological Department, DST Mizoram) in September 2014. It disturbed traffic and blocked the highway for a month in 2014. The slope materials mixed with thick debris and saturates soils. The landslide can be classed as 'debris slide and earth slide' of translational type (Fig. 1). More than 50 lakhs rupees (Approximately USD 78,300) was spend for removal of thick debris to clear the highway (Fig. 1A).

The study area belongs to Upper Bhuban Formation of Surma Group (Miocene age), comprising sandstone, siltstone and thin bedded crumpled shale. The general strike of the area is N065°S and 08° due west. Two sets of joints (J_1: 74°/257° & J_2: 89°/217°) exposed at about 3 m in the western side of the rupture surface, but no continuity observed. Weak sandstone- siltstone bed of shallow marine environment exposed in the crown. An overburden soil, which is regolith/ colluvium type of about 1.5 m thick with very loosely compact in nature. The general slope is about 40° towards east direction in northern side and west in southern side (Figs. 2 and 9). More

S. Hemeda and M. Bouassida (Eds.): GeoMEast 2018, SUCI, pp. 185–193, 2019.
https://doi.org/10.1007/978-3-030-01941-9_16

Fig. 1. Rangvamual landslide along NH-54, Aizawl

than 10 tension cracks are observed in and around crown area, along with seepage in the western side. The corners points of northern and southern area are observed as vulnerable sites (Fig. 1D). For the present study, the stability analyses were performed with the help of numerical program Slide v.6 based on limit equilibrium method (LEM) (Rocscience 2010).

Methodology

(1) ***Soil testing:*** Representative undisturbed soil samples are collected from crown area using core cutter, after dig out surface soils (Fig. 1F). Field dry density and moisture content determined at in-situ using Rapid Moisture Meter (IS: 2720 (Part II). Atterberg limits- Liquid and Plastic limit are determined as per IS: 2720 (Part 5)- 1985 guideline. Represent soil samples are remoulded and performed Direct Shear Test as per IS: 2720 Part 13 guideline.

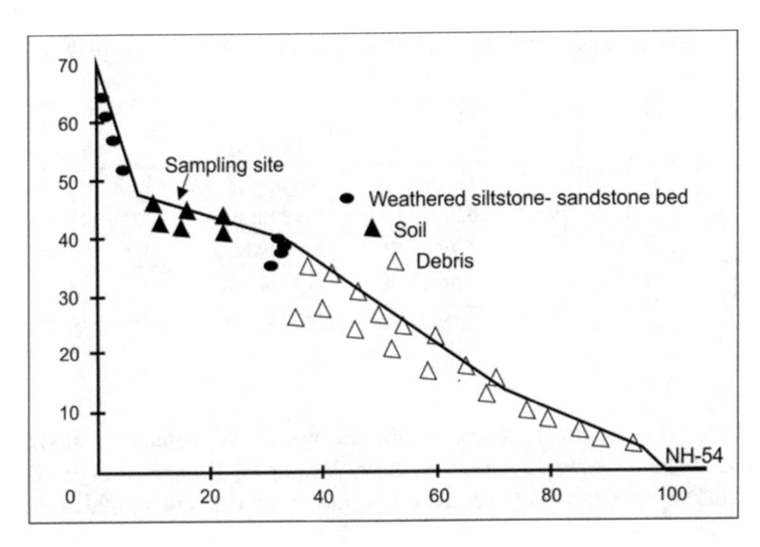

Fig. 2. Slope profile for LEM analysis

(2) ***LEM analysis:*** The study area was analyzed in a LEM package, Slide 6.0 using Ordinary/Fellenius method, Bishop simplified method, and Janbu simplified method, Spencer method, US Corps of Engineer method, GLE/Morgernstern Price method (Fellenius 1936; Bishop 1955; Janbu 1968; Spencer 1967; Corps of Engineer 2003; Morgenstern and Price 1965). Circular slip surfaces with auto refine search method of computation were employed for the stability assessment. In auto refine search, the search for the lowest safety factor is refine as the search progresses. An iterative approach is used, so that the results of one iteration are used to narrow the search area on the slope in the next iteration (Rocscience 2010). It is used to calculate the safety factor of slope based on Mohr- Coulomb criterion (Lin et al. 2014). For LEM of the study area, immediate vulnerable slope in dry condition has been analyzed (Fig. 2).

Results and Discussions

Detailed Geotechnical nature of soils are tested in PWD Quality Control Division, Aizawl and analysed the result using Slide v. 6 software for LEM analysis. Geotechnical parameters such as *angle of friction* (ϕ), *cohesion* (c), *bearing capacity* (q_{ult}) are obtained from Direct Shear Method (IS: 2720 Part XIII), in-situ field dry density and moisture content, and Atterberg's limits are determined.

The observed field density (γ_d) is 16.20 kN/m^3, angle of internal friction 11.31°, cohesion is 4.00 kN/m^2 and ultimate bearing capacity is 220.98 kN/m^2. The observed values of ϕ and cohesion are far less beyond the safety values. The result of atterberg limits are listed in Table 1. Field dry density ranges from 17.35 to 19.40 kN/m^2 and moisture content 10.20% to 13.10% indicated loose and dry soils. 3.79 to 12.18 plasticity index and 1.72 to 3.29 consistency index shows that hard and low plastic characteristics of soils (Table 1). So, the area may activate to large slide after moderate to heavy rainfalls. The area is not suitable for large construction, strengthening and modifying the slope is the utmost needs.

Table 1. Characteristics of Soils [Coduto 1999 (Modified after Sowers 1979) & BS 5930 1999 + A2, 2010]

Sample No.	LL	PL	PI	Classification (*Based on PI*)	c	States of Soil (*Based on* I_c)
RV-1	33.85	21.67	12.18	Slightly plastic	1.72	Medium
RV-2	26.00	19.68	6.32	Low plastic	2.04	Hard
RV-3	22.70	18.91	3.79	Low plastic	3.29	Hard
RV-4	19.70	Nil	Non-plastic	Non-plastic	–	–
RV-5	26.80	19.72	7.08	Low plastic	2.20	Hard

LEM analyses is done in dry conditions based on various methods such as *Ordinary/Felnenius, Bishop Simplified, Janbu Simplified, Spencer, Corps of Engineer #1 and GLE/Morgenstern methods* using parameter such as unit weight, cohesion and angle of friction (Rocscience 2010; Dinpuia 2015). The FS (Factor of Safety) deterministic values from these methods are 0.518, 0.532, 0.516, 0.529, 0.533 and 0.529 respectively. The interpret LEM for various methods are given in Figs. 3, 4, 5, 6, 7 and 8.

Suggestions

Benching and proper drainage system are suggested to decrease driving forces; and to increase resisting forces, construction of different walls with geotextile and vegetation covering are suggested, and are best option to be adopted (After Abramson et al. 2002;

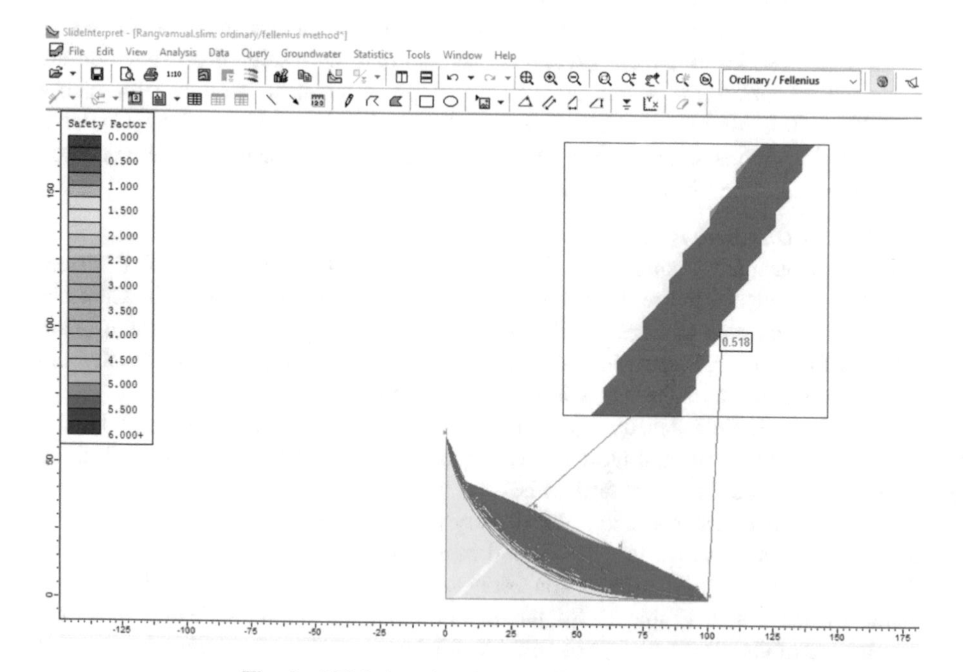

Fig. 3. LEM plot after Ordinary/Fellenius method

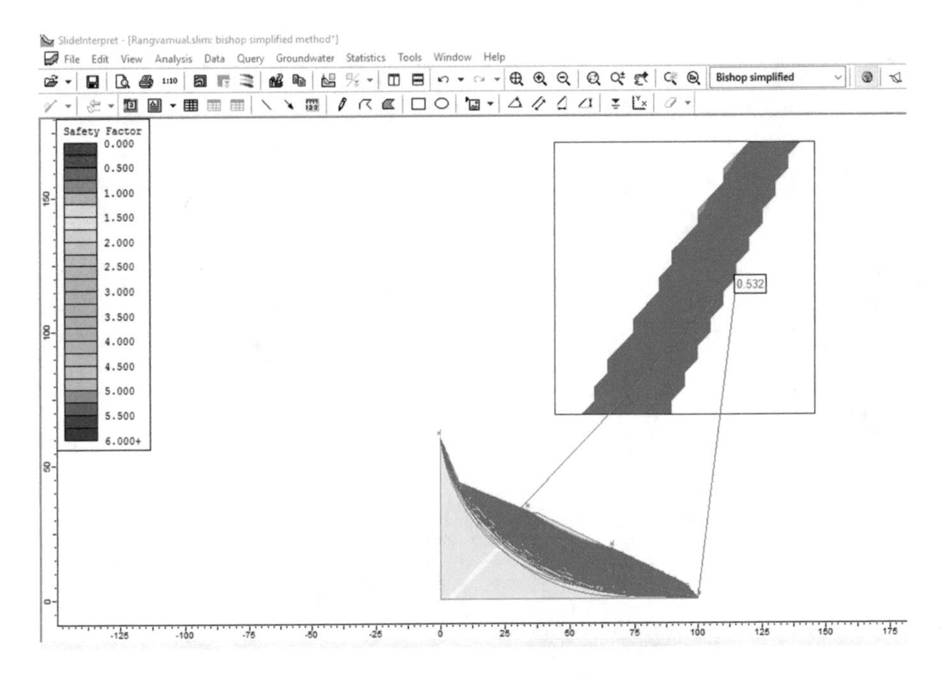

Fig. 4. LEM plot after Bishop simplified method

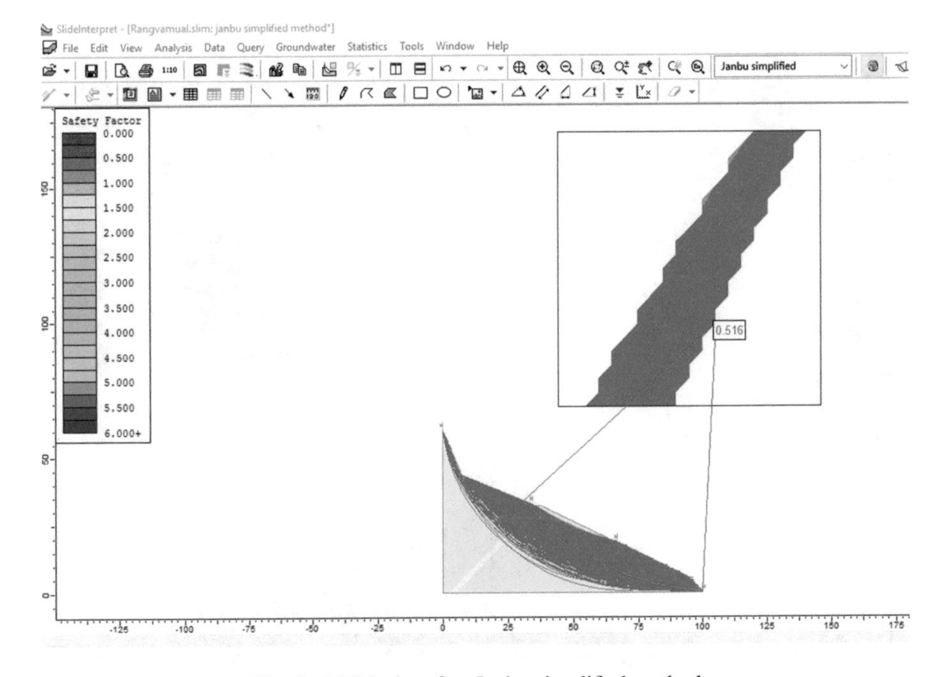

Fig. 5. LEM plot after Janbu simplified method

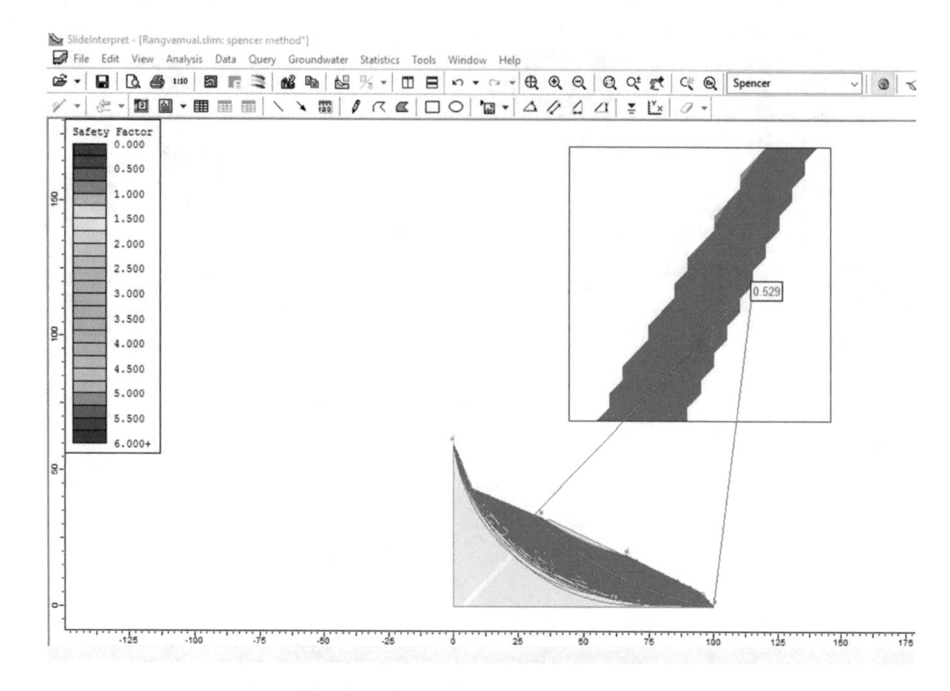

Fig. 6. LEM plot after Spencer method

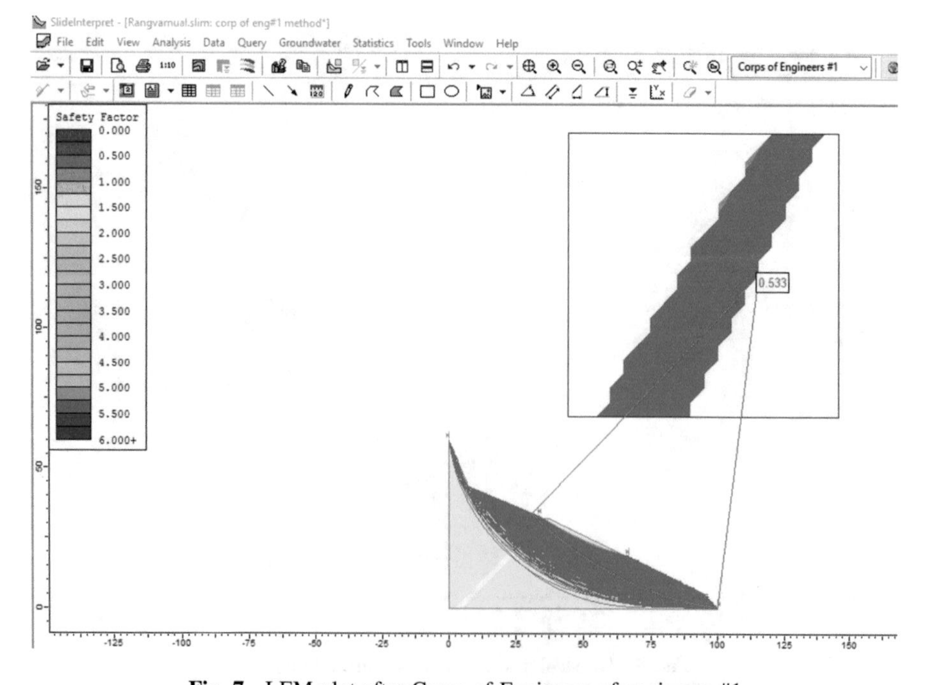

Fig. 7. LEM plot after Corps of Engineers of engineers #1

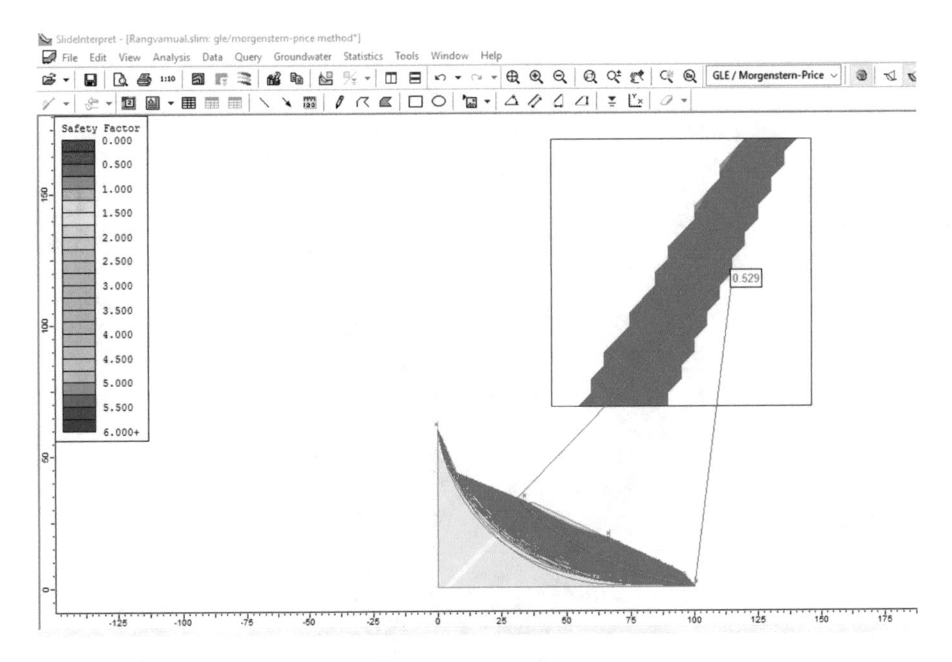

Fig. 8. LEM plot after GLE/Morgenstern method

Baker and Marshall 1958; Duncan 1996; Duncan and Wright 2005; Schuster and Krizek 1978; WP/WLI 2001; Fig. 9). Road diversion also proposed to minimize the risk of various disasters.

(1) **Benching & Gabion Wall:** Two gabion type benching structures are suggested as drawn in Fig. 9 and at least benching on the northern side. Gabion wall with about 3.0 m width and height box at benching. About 15 m length gabion wall (At least 2.0 m width) as breast wall should construct along previous NH in the western section to prevent the movement of tension cracks.

(2) **Toe Wall:** RCC toe walls are suggested at the northern and southern corners along the highway, i.e. highly vulnerable landslip sites. About 20 m length of 3 m height in the eastern site and 30 m length of 3 m height in the western site are proposed. The walls will support the load and increase resisting forces. These walls will be reinforced by anchoring with 25 mm steel, at least 2 m depth. Strong RCC column and beams constructing from toe wall covering the rupture surface.

2 Conclusions

Rangvamual landslide is one of the complex geologic nature. Deterministic analyses (LEM) shows that FS is below unity, i.e. 0.52 and more than 90% probable to slope failure. This may be due to weak cohesion, low angle of friction and loose soils. Various

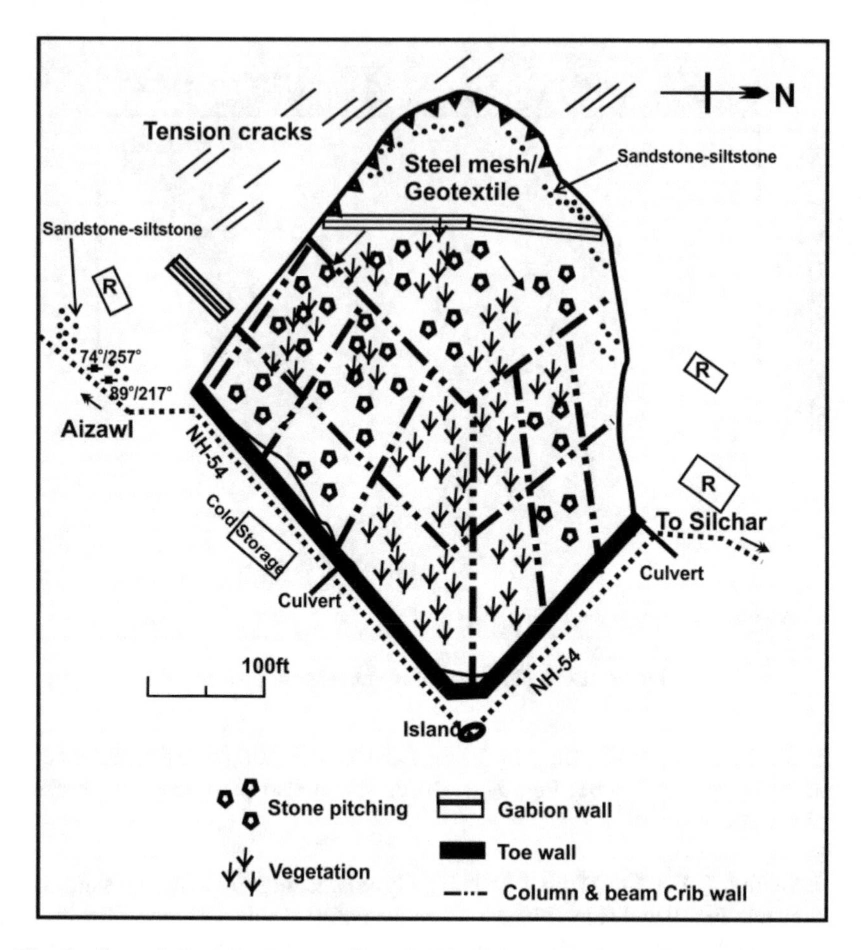

Fig. 9. General Geology & suggestion of mitigation measures of Rangvamual landslide

types of mitigation and protective measures are suggested for the slope stability to minimize the high risk of the State of Mizoram lifeline, NH-54 and airport road.

Acknowledgments. The present study is a part of consultative works with PWD (Highway), Govt. of Mizoram. The author is grateful to the authorities of PWD (Highway) Division for providing laboratory works.

References

Abramson, L.W., Lee, T.S., Sharma, S., Boyce, G.M.: Slope Stability and Stabilization Methods, 2nd edn., pp. 329–461. John Wiley & Sons, New York (2002)

Baker, R.F., Marshall, H.E.: Control and correction. In: Eckel, E.B. (ed.) Landslides and Engineering Practice. Highway Research Board, Special Report No. 29. Committee on Landslide Investigations. NAS-NRC Pub 544, Washington D.C., pp. 150–188 (1958)

BIS, 1985. IS: 2720 (Part II)- 1973. Indian Standard Methods of test for soils, Part V Determination of water content, UDC -131.431.3, 19 p. (2002, reaffirmed)

BIS, 1985. IS: 2720 (Part 5)- 1985. Indian Standard Methods of test for soils, Part V Determination of liquid and plastic limit, UDC 624-131.532-3, 20 p.

BIS, 1994. IS: 2720 (Part 8)- 1983. Indian Standard Methods of test for soils, Part 8 Determination of water content- dry density relation using heavy compaction (Second Revision), UDC 624-131.431.3.624-131.431.5, 19 p.

BIS, 1996. IS: 2720 (Part 13)- 1986. Indian Standard Methods of test for soils, Part 13 Direct Shear Test (Second Revision), UDC 624: 131.439. 5, 12 p.

Bishop, A.W.: The use of the slip circle in the stability analysis of slopes. Geotechnique **5**(1), 7–17 (1955)

BS 5930, 1999 + A2, (2010). Code of practice for site investigations. British Standards Institution

Cheng, Y.M., Lansivaara, T., Wei, W.B.: Two- dimensional slope stability analysis by limit equilibrium and strength reduction methods. Comput. Geotech. **34**, 137–150 (2007)

Coduto, D.P.: Geotechnical Engineering: Principles and Practices, pp. 125–130, 136–155. Prentice Hall, New Jersey (1999)

Duncan, J.M.: Soil slope stability analysis. In: Turner, A.K., Schuster, R.L. (eds.) Landslide, Investigation and Mitigation, Washington D.C., Special report 247, pp. 337–371 (1996)

Duncan, J.M., Wright, S.G.: Soil Strength and Slope Stability, 310 p. John Wiley & Sons, Canada (2005)

Fellenius, W.: Calculation of the stability of earth dams. In: Proceedings of the 2nd Congress on Large Dams, vol. 4. US Government Printing Press, Washington D.C. (1936)

Janbu, N.: Slope stability computations. In: Soil Mechanics and Foundation Engineering Report. Technical University of Norway, Trondheim (1968)

Dinpuia, L.: Geological studies of the rocks at landslide prone localities in Aizawl, Mizoram. Ph. D Thesis, Mizoram University, p. 150 (2015, unpublished)

Lin, H., Zhong, E., Xiong, W., Tang, W.: Slope stability analysis using limit equilibrium method in nonlinear criterion. Sci. World J. **2014**, 7 p. (2014). Article ID 206062. http://dx.doi.org/10.1155/2014/206062

Morgenstern, N.R., Price, V.E.: The analysis of the stability of genreal slip surfaces. Geotechnique **15**(1), 79–93 (1965)

RocScience Inc.: Slide Version 6.0- 2D Limit Equilibrium Slope Stability Analysis, Toronto, Ontario, Canada (2010). www.rocScience.com

Schuster, R.L., Krizek, R.J. (eds.): Landslides Analysis and Control, p. 234. National Academy of Sciences, Washington D.C. (1978). Special Report 176

Singh, P.K., Wasnik, A.B., Kainthola, A., Sazid, M., Singh, T.N.: The stability of road cut cliff face along SH-121: a case study. Nat. Hazards **68**, 497–507 (2013)

Singh, T.N., Singh, R., Singh, B., Sharma, L.K., Singh, B., Ansari, M.K.: Investigations and stability analyses of Pune district, Maharashtra, India. Nat. Hazards **81**, 2019–2030 (2016)

Spencer, E.: A method of analysis of the stbility of embankments assuming parallel interslice forces. Geotechnique **17**(1), 11–26 (1967)

US Army Corps of Engineer: Slope Stability. Engineer manual, No. EM 1110-2-1902, Department of the Army, US Army Corps of Engineer, Washington D.C., 205 p. (2003)

WP/WLI: International geotechnical societies' UNESCO working party on world landslide inventory: a suggested method for reporting landslide remedial measures. Bull. Eng. Geol. Environ. **60**, 69–74 (2001)

Geo-Mechanical Characterization in Laterite Soil Mixtures - Aerial Lime for Road Based Use in Federal District, Brazil

Thamara Silva Barbosa(✉), Lucas Gabriel Lopes da Silva, and Rideci de Jesus da Costa Farias

Catholic University of Brasilia, Brasilia, Brazil
sbthamara@gmail.com, lukkasgabriel0@gmail.com, rideci.reforsolo@gmail.com

Abstract. Soil is widely used in road engineering works. However, it does not always satisfy the specifications for the pavement layers (bases, sub-bases, and subgrade reinforcement) in which it can be used, and it is necessary to look for alternatives so that the paving project is executed. Among the existing alternatives, it is mentioned: to accept the natural soil and adapt the projects to the limitations imposed by it; Remove the material and replace it with a better quality one or adapt the properties of the existing soil in order to create a material capable of meeting the project requirements. The latter is called soil stabilization, which is a cost-effective solution for the preservation of natural resources. For the lateritic soil of the Distrito Federal, it was chosen to perform stabilization with lime, which is an abundant and low cost material in the district. In this presentation, seven lime contents (1%, 2%, 3%, 4%, 5%, 6% and 7%) and two different types of quicklime (CHI and CHIII) were used. The soil characterization tests were carried out and afterwards the Compaction and California Bearing Ratio (CBR) tests of the blends were performed. Results obtained show that there was little variation in moisture content with values ranging between 33.01% and 35.80% for CHI and between 32.10% and 36.21% for CHIII. There was also little variation in dry density, with values ranging between 1.26 g/cm^3 and 1.29 g/cm^3 for CHI and 1.24 g/cm^3 and 1.31 g/cm^3 for CHIII. For values of CBR, there was gradual increase according to the percentage of lime content to the two types of specimens. These values of CBR ranged between 1.42% and 34.63% for CHI lime, and between 2.70% and 10.50% for CHIII lime. With this, it is concluded that, it is more advantageous to stabilize this type of soil with the hydrated lime CHI in order to obtain high values of CBR, which is an indication of the soil compaction characteristics.

1 Introduction

Balbo (2007) affirms that the construction of transport routes is a collective activity from the most remote civilizations, which had its given for reasons of economic order, regional integration and security, and has also been in existence for a long time. The paving of roads, from an antiquity, has become an essential trail for the adaptation and preservation of the most strategic paths.

S. Hemeda and M. Bouassida (Eds.): GeoMEast 2018, SUCI, pp. 194–202, 2019.
https://doi.org/10.1007/978-3-030-01941-9_17

However, a survey carried out by the National Transport Council (2015) shows that, in spite of the outstanding predominance of road transport in the movement of goods and people, it can be seen that, in relation to the total extension of the road network in Brazil (1,691,522 km), only 12% of it is paved. But 42.7% of the total network offer traffic conditions with adequate level of safety and comfort for the users with the other 57.3% present some kind of deficiency in the general state and pose a risk to users who travel on them, and thus urgent for rehabilitation.

With this scenario there is a need to find solutions that are technically possible and economically viable. According to Cristelo (2001), among the various forms studied and used is the stabilization of soils, because in many cases, these soils do not generally meet the requirements for the execution of a road pavement. Thus these soiks are used as they are or in combination whit the following solutions: (a) to accept the natural soil and to adapt the projects to the constraints imposed by it (b) to remove the local material while replacing it with a better quality one and (c) to adapt the properties of the existing soil in order to create a material capable of meeting the design requirements.

Senço (2001) opined that this type of stabilization can give an economical solution and better technique, when compared to the labour in removing the clay soil in order to substitute it with a granular soil. Among the benefits of using this method, the highlight of the improvement of the soil properties is, mainly in terms of the ability to support and reduce soil shrinkage, swelling or expansion. Therefore, it is the general objective to characterize this soil and the soil-lime mixtures in pre-established percentages.

According to Metogo (2010), the use of lime for the chemical stabilization of soil is important to improve the physical and mechanical properties of this, with the possibility of being used in road projects. It is common to use virgin lime and also hydrated in practice. For the stabilization of the soil only few amounts of lime are necessary, because its main purpose is the gain of resistance and decrease of plasticity. However, Palmer (1986) reports that hydrated quays are more stable than virgins.

Azêvedo (2010) says that in relation to the use of soil stabilization with lime in Brazil, there are a few experimental excerpts dating mainly from the 1970s and some point solutions from the 1990s and 2000s. Despite the good results presented in the soil stabilization experiments with hydrated lime, the number of effective lime applications in this technique in Brazil is not yet in agreement with its potentiality, mainly due to the lack of knowledge of the lime action in the soil, physical and chemical characteristics of the materials used by the system and the methodology of the execution details of the work.

Whitin this context, the main objective of this research is to encourage the use of hydrated lime as a stabilizer in road works, since, besides being a cheap material, easy to find and handle, in a few days there is already a result expressive in the soil resistance gain as a function of the percentage used without having to wait long days for the cure.

For this purpose, the specific objectives are to analyze the behavior of the hydrated lime mixtures, CHI and CHIII, as applied to the laterictic soil and to obtain the optimum content or percentages that is feasible and suitable for use as base material in road pavements.

1.1 Methodology

In this work, laboratory tests were carried in order to determine the physical and mechanical characteristics of soil and soil-lime mixtures. Soil samples were collected at the Universidade Católica of Brasília campus in Águas Claras and the CHI and CHIII wharves that were used followed the standards established by NBR 7175 (2002), according to the manufacturers' information.

Based on the assumption that the effect of lime on the lateritic soil of Brasilia is not known, since the chemical reactions in this may vary according to its composition and the organic matter content, and with the objective of evaluating the resistance gain in a short time of cure and according to the increment to every 1% of lime, it was chosen to use seven percentages of lime (1%, 2%, 3%,4%, 5%, 6% and 7%) were employed for each type of lime used. According to Balbo (2007), the soil-lime mixture (SCA) dosage criteria vary according to the intended use for the blend. When used as a subgrade reinforcement material or as a subbase, the dosage can be performed using the CBR criterion (for CBR greater than 20%). The viability of these mixtures was therefore evaluate from the results of the CBR assays.

Preparation of Samples
After the collection of the disturbed sample of soil, the preparation of test specimens was started for the characterization tests; for this, ABNT NBR 6457 (1986) was used as the code.

Granulometric Analysis
According to Senço (2007), the granulometric analysis of a soil allows the knowledge of the percentages of the particles that constitute it as a function of its dimensions. This knowledge is of great importance for studies of soil behavior both as a foundation element on which the pavement is supported and as the various layers of the pavement.

The granulometry test carried out followed the guidelines of ABNT NBR 7181 (1982) by combining sieving and sedimentation in a 24 h interval.

Specific Mass
According to Senço (2007), the specific mass can be real or apparent according to the volume considered. The actual specific mass is related to the mineralogical nature and is the relationship between the mass of the solid part and the volume of solids; the apparent specific mass of the wet soil is the ratio of the wet soil mass to the volume occupied by it and the apparent dry mass of the dry soil is the ratio of the mass of solids to the total volume.

Determination of the specific mass of the solids was executed according to the procedure described in ABNT NBR 6508 (1984).

Atterberg Boundaries
Caputo (1988) defines Liquidity Limit as the boundary between the liquid state, whose soil moisture is very high and this is presented as a dense fluid, and the plastic state, where the soil loses its ability to flow, but can be molded easily, while retaining its shape.

The Plasticity Limit is the boundary between the plastic and the semi-solid state, in which the moisture loss continues and the soil is destroyed when it is woked upon.

Caputo (1988) also defines the Plasticity Index as the difference between the Liquid Limit (LL) and the Plasticity Limit (PL).

Table 1 presents the soil classification proposed by Jenkins and Caputo (1988) according to the value of the Plasticity Index (PI).

Table 1. Soil classification according to IP

IP	Classificação
$1 < IP < 7$	Fracamente plásticos
$7 < IP < 15$	Medianamente plásticos
$IP > 15$	Altamente plásticos

The Liquidity and Plasticity Limit tests were based on the specifications ABNT NBR 6459 (1984) and ABNT NBR 7180 (1984), respectively.

Compaction
According to Senço (2007), the soil compaction is the operation that reduces the voids in the soil, while compacting it by mechanical means.

The compaction for obtaining the optimum moisture and soil specific mass was based on the recommendation of ABNT NBR 7182 (1986). The normal compaction energy (five layers of twelve strokes in each layer) was used in the test.

California Bearing Ratio (CBR)
According to Bernucci et al. (2006), the resistance in the CBR test indirectly combines the cohesion and the friction angle of the material. The CBR is defined as the ratio between the pressure required to produce a piston penetration in a soil test piece or granular material and the pressure required to produce the same penetration in the reference standard material, the result is expressed as a percentage.

The CBR tests were performed according to ABNT NBR 9895 (1987) recommendations. The normal energy was used as in the compaction test.

1.2 Results

Granulometric Analysis
From the test results which are plotted in Fig. 1, and while using the code it noticed that with the addition of deflocculant, the percentage of the material passing through the clay classification is higher when compared to the percentage of clay of the material that was added whit only water. The deflocculant, therefore, is efficient with respect to the separation of the grains which remained together after the manual dewatering.

Likewise, the percentage of sand (fine, medium or coarse) is lower when it comes to the addition of water in relation to the addition of hexametaphosphate, since some of the material presented as small clods that were retained in the sieves of sand classification or fraction.

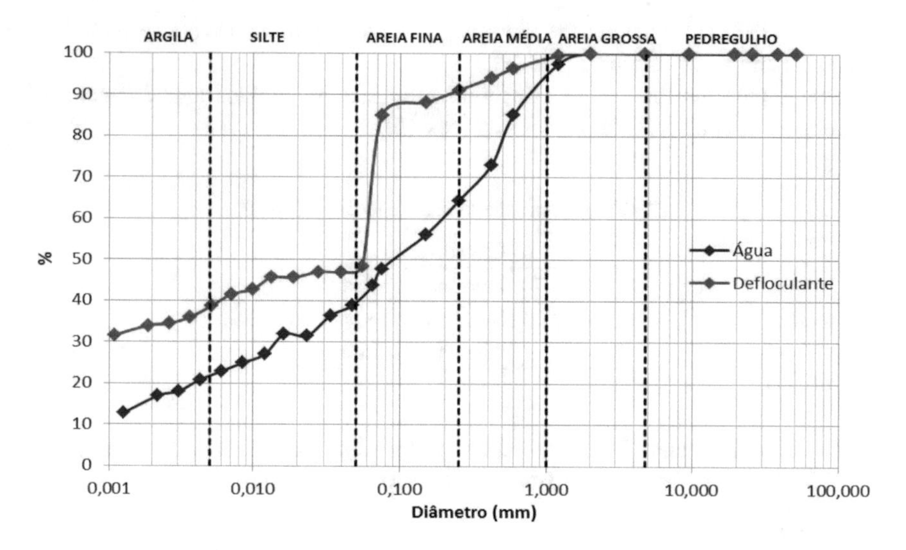

Fig. 1. Graph of the sedimentation granulometry test

Specific Mass
After the test to determine the specific mass of the soil, the result was 2.61 g/cm^3

Atterberg Boundaries
After the Liquidity Limit test was carried out, according to the recommendations specified in the code, values of moisture content in the soil are as shown in Table 2. The average of these values corresponds to the value of LL = 48.80%.

Table 2. Result of the Liquidity Limit test

Determination	Number of strokes	Capsule	Tara (g)	Tara + Soil (g)	Tara + Solid (g)	Humidity (%)
1	24	18	18,60	22,33	21,11	48,61
2	31	276	17,17	20,18	19,23	46,12
3	35	440	13,27	17,57	16,19	47,26
4	21	306	16,25	19,80	18,61	50,42
5	16	16	18,69	22,83	21,41	52,21

Also from the Liquidity Limit test, according to the corresponding standard, the values of moisture content humidity were obtained as shown in Table 3. The average of these values corresponds to the value of Plastic Limit, PL = 34.38%.

From Eq. (1), the Plasticity Index given by PI = 14.42% as calculated. It can be Seen from the data displayed in Table 1, that the soil can be classified as that with Medium Pasticity.

Table 3. Result of the Plasticity Limit assay

Determination	Number of strokes	Tara (g)	Tara + Soil (g)	Tara + Solid (g)	Humidity (%)
1	385	14,52	15,88	15,54	33,33
2	407	13,98	15,78	15,30	36,36
3	264	17,97	19,33	18,97	36,00
4	313	20,41	21,84	21,48	33,64
5	316	16,76	17,73	17,48	34,72
6	358	14,29	15,07	14,88	32,20

Compaction

After compaction of the natural soil and soil-lime mixtures, the compaction were plotted from the results obtained, according to the specification of the code for this test. From the curves, the data of Optimum Humidity or Optimum Moisture Content and Dry Specific Mass are as shown in Table 4.

Table 4. Compaction assay results

Material	Great humidity (%)	Specific dry mass (g/cm^3)
Natural Soil	33,85	1,31
Soil + 1% de CHI	35,80	1,27
Soil + 2% de CHI	34,38	1,29
Soil + 3% de CHI	34,62	1,28
Soil + 4% de CHI	34,24	1,27
Soil + 5% de CHI	35,09	1,26
Soil + 6% de CHI	33,01	1,26
Soil + 7% de CHI	34,53	1,26
Soil + 1% de CHIII	32,93	1,30
Soil + 2% de CHIII	33,33	1,31
Soil + 3% de CHIII	32,10	1,31
Soil + 4% de CHIII	33,51	1,28
Soil + 5% de CHIII	36,21	1,24
Soil + 6% de CHIII	34,91	1,27
Soil + 7% de CHIII	35,78	1,27

California Bearing Ratio (CBR)

The CBR test was performed according to the standard specifications for this test. The results obtained from Expansion and CBR of the natural soil and the soil-lime mixtures are shown in Table 5.

According to the DNIT Pavement Manual (2006), the subgrade materials should have an expansion, measured in the ISC test, of less than or equal to 2% and an CBR $\geq$ 2%; the materials for reinforcement of the subgrade, those with CBR greater

Table 5. ISC test results

Material	Expansion (%)	ISC (%)
Soil natural	0,44	7,63
Soil + 1% de CHI	0,09	1,42
Soil + 2% de CHI	0,09	5,77
Soil + 3% de CHI	0,01	14,90
Soil + 4% de CHI	0,01	18,65
Soil + 5% de CHI	0,01	21,29
Soil + 6% de CHI	0,01	27,48
Soil + 7% de CHI	0,04	34,63
Soil + 1% de CHIII	0,09	2,70
Soil + 2% de CHIII	0,01	3,41
Soil + 3% de CHIII	0,01	5,39
Soil + 4% de CHIII	0,01	5,71
Soil + 5% de CHIII	0,01	7,71
Soil + 6% de CHIII	0,01	9,66
Soil + 7% de CHIII	0,01	10,50

than that of the code and expansion $\leq 1\%$; the materials for sub-base, those with ISC $\geq$ 20% and expansion $\leq 1\%$ and materials for base, which have: CBR $\geq$ 80% and expansion $\leq$ 0.5%, Liquidity limit $\leq$ 25% and Plasticity Index $\leq$ 6%. Therefore, the soil with a 1% CHI does not meet any of the criteria, the percentages of 5%, 6% and 7% of CHI can serve as sub-base material and the other percentages meet the criteria for subsoil materials and strengthening of subgrade.

2 Conclusions

By analyzing the soil particle size test and the plasticity index, the soil used in this work can be classified as clay-silt soil of medium plasticity (ML or OL) according to USCS. According to Balbo (2007), the apparent dry bulk density of a soil-lime mixture is lower than that of the natural soil (accentuated by the increase in lime consumption) and the optimum moisture content is increased after the soil was mixed with lime. This trend can be observed in the results of the compaction test presented in Table 4.

However, as the percentage of lime increases there was also a corresponding variation in the results obtained, and in conjunction to factors such as that the test was performed in an environment without a controlled temperature and Brasília presented a wide thermal variation during the day which may thus interfere with the result of the compaction test. In relation to the California Support Index (ISC) test, there was no expansion related standard, but for all mixtures, the results obtained meet the standards established by DNIT. It was observed in this study that Hydrated lime I (CHI) presents better ISC results when compared to the same percentages of Hydrated lime II (CHII) and that according to the percentage of lime used, the value of ISC. However, none of the percentages of the soil-lime mixtures meet the basic material criterion established by the DNIT. It is possible that the type of lime used exerts a certain influence on the results of the tests, since CHI is purer than CHIII because it has a higher amount of calcium hydroxide as specified in NBR 7175/2002, which may interfere in the binding capacity of lime.

Considering the percentages of lime used, it is possible that there is an influence on the final cost of the soil-lime mixture, however, the research focused on the possibility of not having a deposit available in the region and with that having to use a larger amount of lime in the mix to improve soil stability.

References

ABNT – Associação Brasileira de Normas Técnicas. NBR 6457. Amostras de solo – Preparação para ensaios de compactação e ensaios de caracterização. Rio de Janeiro (1986)

ABNT – Associação Brasileira de Normas Técnicas. NBR 6459. Solo – Determinação do limite de liquidez. Rio de Janeiro (1984)

ABNT – Associação Brasileira de Normas Técnicas. NBR 6508. Grãos de solo que passam na peneira de 4, 8 mm – Determinação da massa específica. Rio de Janeiro (1984)

ABNT – Associação Brasileira de Normas Técnicas. NBR 7180. Solo – Determinaçaõ do limite de plasticidade. Rio de Janeiro (1984)

ABNT – Associação Brasileira de Normas Técnicas. NBR 7181. Solo – Análise granulométrica. Rio de Janeiro (1984)

ABNT – Associação Brasileira de Normas Técnicas. NBR 7182. Solo – Ensaio de compactação. Rio de Janeiro (1986)

ABNT – Associação Brasileira de Normas Técnicas. NBR 7175. Cal hidratada para argamassas – Requisitos. Rio de Janeiro (2002)

ABNT – Associação Brasileira de Normas Técnicas. NBR 9895. Solo – Índice de suporte Califórnia. Rio de Janeiro (1987)

Azêvedo, A.L.C.: Estabilização de solos com adição de cal – Um estudo a respeito da reversibilidade das reações que acontecem no solo após a adição de cal. Masters dissertation – Universidade Federal de Outro Preto, Escola de Minas (NUGEO) (2010). CDU: 624.12:691.51<Query ID="Q5" Text="Please check and confirm the year for Ref. “Azêvedo (2010)”. –>

Balbo, J.T.: Pavimentação asfáltica: materiais, projetos e restauração, 558 p. Oficina de Textos, São Paulo (2007). ISBN 978-85-86238-56-7

Bernucci, L.B., et al.: Pavimentação asfáltica: formação básica para engenheiros, 504 p. Petrobras, Abeda, Rio de Janeiro (2006). ISBN 85-85227-84-2

Caputo, H.P.: Mecânica dos Solos e suas aplicações, 234 p. LTC, Rio de Janeiro (1988). ISBN 978-85-216-0559-1

Confederação Nacional Do Transporte: Pesquisa CNT de rodovias 2015. http://www.cnt.org.br/Imprensa/noticia/pesquisa-cnt-de-rodovias-2015-maior-parte-dos-trechos-apresenta-problemas. Acesso em 10 fev 2017

Cristelo, N.M.C.: Estabilização de solos residuais graníticos através da adição de cal. Dissertação (Mestrado em Engenharia Civil) – Escola de Engenharia da Universidade do Minho (2001)

DNIT – Departamento Nacional de Infraestrutura de Transporte. Manual de pavimentação. 3. edn. DNIT, Rio de Janeiro (2006)

Metogo, D.A.N.: Construção e avaliação inicial de um trecho de pavimento asfáltico executado com misturas de solo tropical, fosfogesso e cal. Masters dissertation. Universidade Federal de Góias, Escola de Engenharia Civil (2010). CDU: 631.445.7

Palmer, C.: Virginia's lime industry. Virginia Mineral, United State of America, vol. 32(2), pp. 33–44, Nov 1986

de Senço, W.: Manual de técnicas de pavimentação, vol. I, 2nd edn, 671 p. Pini, São Paulo (2007). ISBN 978-85-7266-199-7

de Senço, W.: Manual de técnicas de pavimentação, vol. II, 671 p. Pini, São Paulo (2001). ISBN 85-7266-125-5

Effects of Sand Sizes on Engineering Properties of Tropical Sand Matrix Soils

Aminaton Marto[1,2] and Bakhtiar Affandy Othman[1,2(✉)]

[1] Disaster Preparedness and Prevention Centre,
Malaysia-Japan International Institute of Technology,
Universiti Teknologi Malaysia, 54100 Kuala Lumpur, Malaysia
aminaton@utm.my, baffandy2@live.utm.my
[2] Centre of Tropical Geoengineering (GEOTROPIK),
Faculty of Civil Engineering, Universiti Teknologi Malaysia (UTM),
81310 Skudai, Johor, Malaysia

Abstract. This paper presents an experimental study focusing on the effects of various sizes of sand on the engineering properties of tropical sand matrix soils, particularly the undrained shear strength. Static triaxial tests on reconstituted samples of sand with 0, 10, 20, 30, and 40% of low plasticity fines content by weight were carried out using GDS ELDYN® triaxial machine. The tests were performed on tropical specimens of three different sizes of sand (coarse, medium, and fine sand) which were mixed with kaolin as the fines content. Samples were tested at 15% relative density under two effective confining pressures of 100 kPa and 200 kPa, respectively. From the results of Consolidated Undrained Triaxial tests, the Critical State Line of the sand matrix soils with different sizes of sand had been developed. Based on the results from stress path diagram, the critical state parameters of sand matrix soils, represented by the critical stress ratio, M, are found to range from 1.41 to 1.35 for coarse, 1.38 to 1.29 for medium, and 1.30 to 1.25 for fine sand matrix soils. The maximum particle density was achieved at lower values of fines content for medium and fine sand matrix soils compared to coarse sand matrix soil. The sand size affects the maximum and minimum void ratio of sand matrix soils. At the same fines content, the void ratio of sand matrix soils increased as the sand size decreased.

1 Introduction

Naturally, soils contain an amount of fines content such as silt and clay or silt/clay. The presence of silt/clay was seen contributing to a geotechnical field problem such as in liquefaction phenomenon. Currently, research on soil liquefaction engineering has been mandatory, which focus on the liquefaction resistance of sand matrix soils and sand containing limiting fines content. The interaction between sand and fines particles appears to be a well-studied subject because the composition characteristic of sand matrix soils is complex.

Numerous researches have been carried out to investigate on how fines particles influence the liquefaction resistance of sand matrix soils, but with contradictory findings. Tan et al. (2013) summarized findings on liquefied soils that occurred at different

S. Hemeda and M. Bouassida (Eds.): GeoMEast 2018, SUCI, pp. 203–213, 2019.
https://doi.org/10.1007/978-3-030-01941-9_18

condition of fines content of 20% to 70%. The liquefaction resistance of sand matrix soils could be increased (Park and Kim 2013; Benghalia et al. 2015) or decreased (Wang and Wang 2010; Monkul and Yamamuro 2011) as the fines content increased. This contradiction might be caused by the biased way of previous research trends that only look into the roles of fines on liquefaction resistance of sand matrix soils. From the results of previous research carried out by Marto et al. (2014), it shows that the concept of threshold fines content explained the effect of fines content on the compositional interaction of sand matrix soils. Although these research findings are in agreement with Lade and Yamamuro (1997), they are not conclusive enough to be applied on all types of sand matrix soils that have different physical characteristics of sand.

Widely known by geotechnical engineer, the slope of deviator stress, q versus mean effective stress p', which is represented by M, has an affect on the behavior of soils (Phan et al. 2016). In 1936, initial research on critical void ratio was carried out by Casagrande which contributed to the development of the Critical State Soil Mechanics (CSSM). Been et al. (1991) defined the state of sand as the physical conditions under which it exists. They also stated that two primary state variables of soils are stress and void ratio. Figure 1 shows the concepts of the critical state (CS) and critical state line (CSL) in stress path ($p' - q$) spaces. The CSL is described in the following equations:

$$e_{cs} = \Gamma_{cs} - \lambda \ln p' \quad (1)$$

$$q = M_{cs} p' \quad (2)$$

where, e_{cs} is the void ratio at critical state under mean effective stress, p'; q is the mean effective stress; M_{cs} or sometimes just written as M is the gradient of the critical state line; and Γ_{cs} is the intercept of the critical state line.

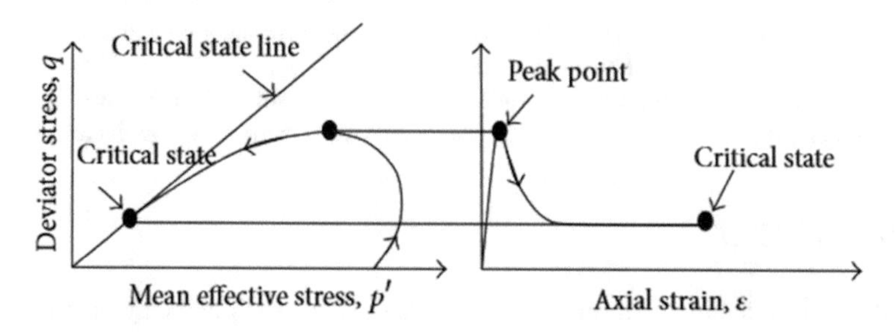

Fig. 1. Undrained behavior of loose soil in stress path diagram (Been and Jefferies 2004)

Lade and Yamamuro (1997) shows the particle interaction of silty sand before and after application of force as shown in Fig. 2. Recent study related to sand fine mixtures on critical state and state parameters was conducted by Phan et al. (2016). Phan et al. (2016) found that liquefaction resistance decreased when silt content increased to a minimal value, then liquefaction resistance increased back as silt content increased. From undrained triaxial test, Phan et al. (2016) showed that the gradient of the critical state line, M decreased from 0.59 to 0.22 as the fine content increases from 0% to 50%. Similar to Phan et al. (2016), Marto et al. (2016) showed that the M decreased as fine content increased from 0% to 25% and then reversely increased until 40% fines content. According to Phan et al. (2016) and Marto et al. (2016), this behaviour occurred due to fully occupied of fines between the sand grains which contributed to a grain separation at certain amount of fines.

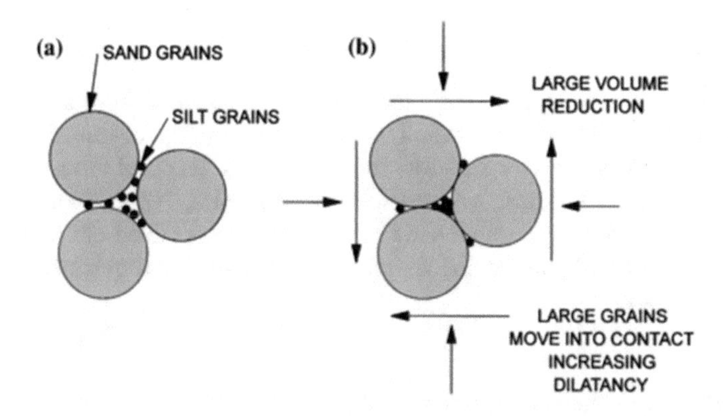

Fig. 2. Particle structure of silty sand before and after loading stress (Lade and Yamamuro 1997)

Results from Nevada 50/200 sand, Nevada 50/80 sand, Ottawa 50/200 sand, and Ottawa F-95 sand conducted by Lade and Yamamuro (1997) show that the static liquefaction potential increased with the increase of fines content at the same low confining pressure, 25 kPa. However, Ottawa 50/200 sand shows higher liquefaction resistance compared with Nevada 50/200 sand. The study from Monkul et al. (2016), who used sand silt mixtures sample, clearly showed that the coarser the base sand gradation, the more liquefiable potential has been monitored, which showed the significant effect of grain size distributions on static liquefaction. In other words, the more gap graded of the soils, the higher potential liquefaction can be monitored.

To date, many empirical evidences reveal that the early assumption that Malaysia is immune from earthquakes seems misleading (Marto et al. 2013). The seismic risk is expected to continue until the next decade because of projected economic and population growth. This challenge is affecting local construction trades and is threatening the ability to construct quality and sustainable structures. The local construction industry must take into consideration the seismic aspect in the future building design code, as indicated by Malaysia Minister of Housing and Local Government Ministry (The Star 2012). Hence, the study on sand matrix soils, which aimed to investigate the effect of different sizes of sand on engineering properties in tropical sand matrix soils, was conducted.

To achieve the aim of this study, the sand was separated according to different sizes: coarse sands (retain 2.0 mm to 0.6 mm), medium sands (retain 0.425 mm to

0.212 mm), and fine sands (retain 0.150 mm to 0.063 mm) which are mixed with kaolin (fines content) at different percentage (0% to 40%) by weight to form the reconstituted sand matrix soils. Static triaxial test was conducted on the sand matrix soils to evaluate and determine the behaviour and engineering properties of tested samples on the results of CSL.

2 Experimental Program

2.1 General

The selected tropical clean sand used for testing was obtained from a mining site at Johor area located 350 km from Kuala Lumpur, Malaysia. The colour of clean sand is white and it is widely used for construction. Three different sand sizes of clean sand were prepared and mixed with white kaolin at 0%, 10%, 20%, 30%, and 40% by weight. A basic property test on all sand matrix samples as tabled out in Table 1 was performed that involved particle size distribution, minimum and maximum void ratio, and particle density.

Table 1. Sample codes and the percentage of sand and kaolin to form the sand matrix soils of different sand sizes

Sand matrix soils	Sample code	Percentage by weight (%)	
		Sand	Kaolin
Clean sand	S100K0	100	0
Coarse sand	S100K0-C	100	0
Coarse sand + Kaolin	S90K10-C	90	10
	S80K20-C	80	20
	S70K30-C	70	30
	S60K40-C	60	40
Medium sand	S100K0-M	100	0
Medium sand + Kaolin	S90K10-M	90	10
	S80K20-M	80	20
	S70K30-M	70	30
	S60K40-M	60	40
Fine sand	S100K0-F	100	0
Fine sand + Kaolin	S90K10-F	90	10
	S80K20-F	80	20
	S70K30-F	70	30
	S60K40-F	60	40

The main experimental task in this study consisted of performing isotropically consolidated undrained (CU) triaxial test on reconstituted samples of sand matric soil under monotonic loading using Enterprise Level Dynamic Triaxial System (ELDYN). As shown in Fig. 3, this equipment, which is located at Geotechnical Engineering Laboratory, Faculty of Civil Engineering, Universiti Teknologi Malaysia, is manufactured by Geotechnical Digital System (GDS), UK. The triaxial tests were conducted in accordance to Part 1 of 3 of Introduction to Triaxial Testing by GDS Instruments which is based on British Standard BS1377: Part 8: 1990.

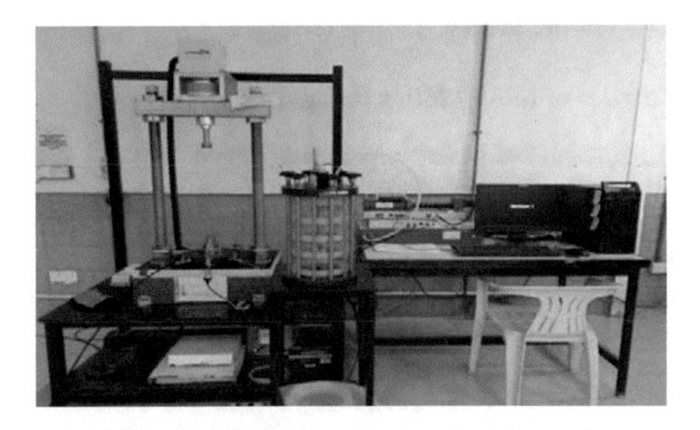

Fig. 3. Enterprise Level Dynamic Triaxial System (ELDYN) machine

2.2 Isotropically Consolidated Undrained Triaxial Test

Marto et al. (2016) observed that the segregation condition of sand matrix soils will lower the resistance of sand matrix soils to liquefy. Hence, segregation of sample, which is the separation of the finer particle and coarser particle that usually occurred at the bottom or at the top of the sample, should be avoided. Past studies carried out by Yilmaz et al. (2008), Chang and Hong (2008), Ibrahim et al. (2010), Takch et al. (2016) and Papadopoulou and Tika (2016) showed that moist or wet tamping technique during sample preparation produced homogenous samples.

In this study, the cylindrical reconstituted samples of 50 mm diameter and 100 mm height were prepared on the base pedestal of triaxial test equipment. In order to ensure the consistency and homogeneous density of the cylindrical samples, the moist tamping technique was applied with 5% of moisture content and 15% relative density. Since granular soils were used for this experiment, a split-part mould was required during compaction of the soil on the pedestal. Once the sample was prepared, then Perspex cell was mounted, and saturation process was conducted to ensure all voids are filled with water. As required by Rees (2013), in order to assist the samples to achieve full saturation, it was necessary to apply de-aired water. Through saturation process, constant effective stress was maintained at about 10 kPa (linear increase of cell pressure and back pressure) until Skempton's B-value achieved was ≥ 0.95.

Each of the soil samples was isotropically consolidated at two effective confining pressures of 100 kPa and 200 kPa. The sign of complete consolidation stage occurred when the back-volume change is observed to have less than 5mm3 over a period of 5 min or the volumetric change remains invariable in 5 min, as stated by Wang and Wang (2010). Once the complete consolidation stage was achieved, the sample was compressed vertically with 0.1 mm/min strain rate, which is acceptable to obtain a steady-state condition of each of the samples, and it continued to shear until failure or up to 25% axial strain.

3 Results and Discussions

3.1 Basic Properties of Sand Matrix Soils

Table 2 lists out the results of basic properties of tested various composition of sand matrix soils. Soil classification of various samples of sand matrix soils are based on the Unified Soil Classification System (USCS). Similar trends were observed for all three sand sizes. The first two samples which consist of 0% and 10% of fines are poorly graded (SP) while all other samples from 20% to 40% of fines content are classified as silty sand (SM). The range of coefficient of uniformity, Cu of coarse, medium, and fine sand matrix soils is 1.71 < Cu < 90.0, 1.60 < Cu < 32.0, and 0.90 < Cu < 7.14, respectively. The coefficient of curvature, Cc of each size of sand matrix soils was between 1.54 < Cc < 17.04 for coarse sand, 0.90 < Cc < 7.14 for medium sand, and 1.03 < Cc < 4.27 for fine sand.

Table 2. Basic properties of sand matrix soils

Code	Density, ρ_s Mg/m^3	Void ratio, e		Gradation		Soil classification
		e_{min}	e_{max}	C_u	C_c	
S100K0-C	2.60	0.87	1.03	1.71	0.96	SP
S90K10-C	2.61	0.81	0.88	7.67	3.71	SP
S80K20-C	2.63	0.73	0.79	46.00	17.04	SM
S70K30-C	2.62	0.69	0.76	73.33	1.36	SM
S60K40-C	2.64	0.73	0.87	90.00	0.54	SM
S100K0-M	2.60	0.90	1.05	1.60	0.90	SP
S90K10-M	2.61	0.84	0.96	2.47	1.62	SP
S80K20-M	2.63	0.82	0.89	14.00	7.14	SM
S70K30-M	2.64	0.80	0.90	32.00	7.03	SM
S60K40-M	2.64	0.86	1.06	30.00	2.13	SM
S100K0-F	2.60	0.99	1.08	2.22	1.25	SP
S90K10-F	2.60	0.92	1.00	2.86	1.03	SP
S80K20-F	2.63	0.98	1.08	12.00	3.70	SM
S70K30-F	2.64	0.99	1.09	15.00	4.27	SM
S60K40-F	2.64	1.06	1.09	13.00	3.25	SM

For the particle density (ρ_s), different sizes of sand is observed to have no significant influence on the value at the same fines content. However, the ρ_s maximum value was obtained at 40% fines content for coarse sand and 30% for both the medium and fine sand matrix soils. Hence, the maximum particle density was achieved at lower values of fines content for medium and fine sand matrix soils compared to coarse sand matrix soil. This is due to the amount of voids which differs between different sizes of sand. Besides that, the minimum (e_{min}) and maximum (e_{max}) void ratio increases with decreases of sand size. At the same fines content, the values increased as the sand size decreased. As an example, at 40% fines content, the e_{min} for coarse, medium, and fine sand matrix soils are 0.73, 0.86, and 1.06, respectively while for e_{max} the values are 0.87, 1.06, and 1.09, respectively.

3.2 Monotonic Triaxial Test

Three sets of monotonic results of different types of sand matrix soils, coarse, medium, and fine sand which were mixed with different percentage of kaolin (0%, 10%, 20%,

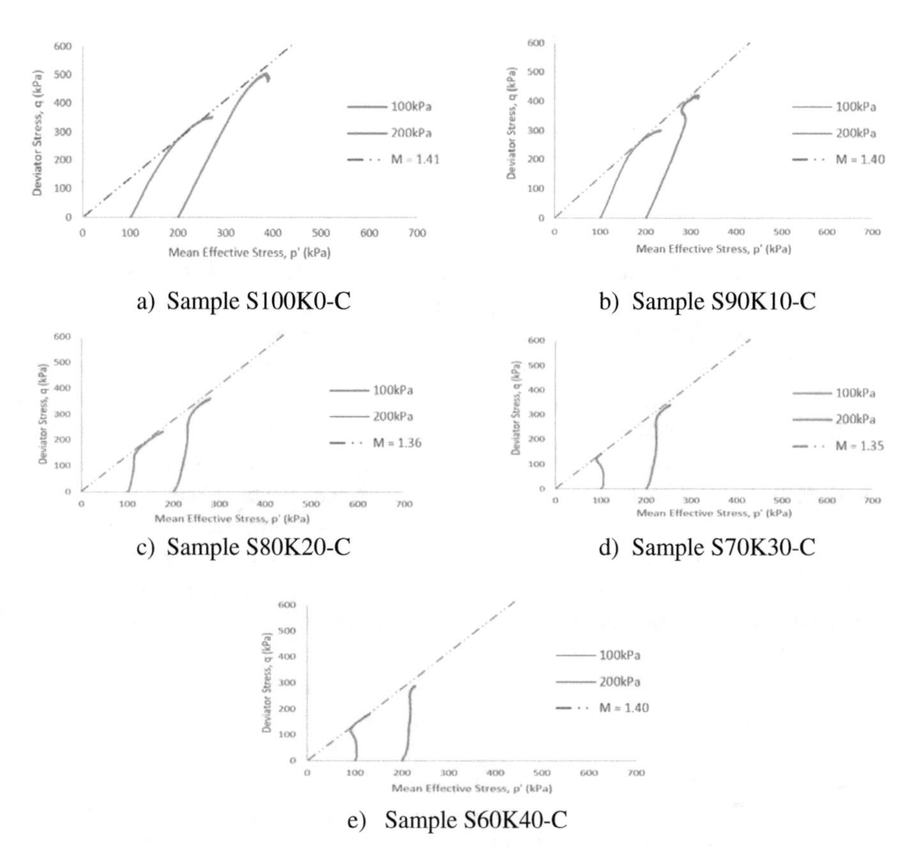

a) Sample S100K0-C

b) Sample S90K10-C

c) Sample S80K20-C

d) Sample S70K30-C

e) Sample S60K40-C

Fig. 4. Relationship between deviator stress and mean effective stress of coarse sand matrix soils

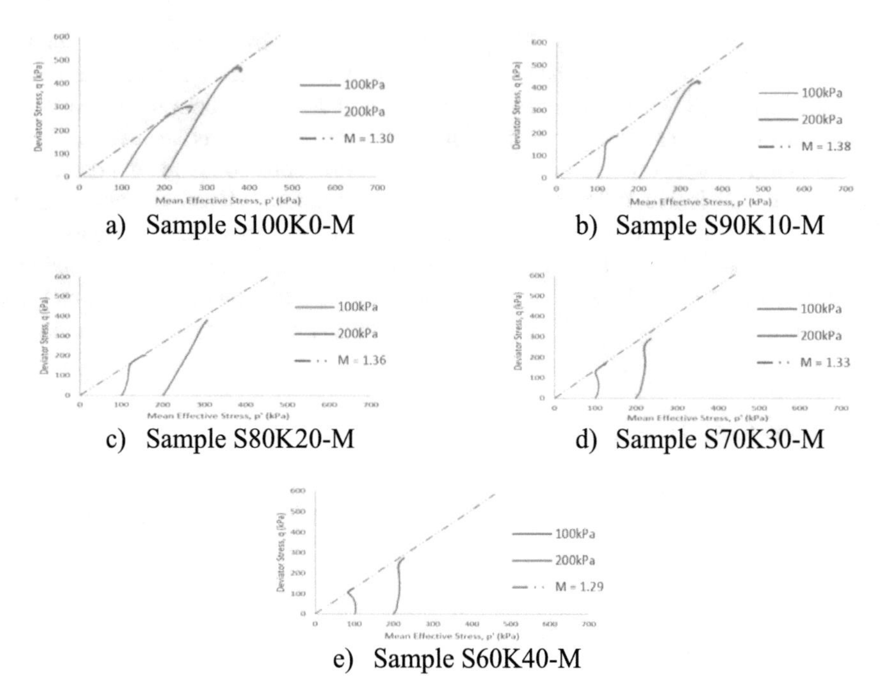

Fig. 5. Relationship between deviator stress and mean effective stress of medium sand matrix soils

30%, and 40%), are illustrated in Figs. 4, 5 and Fig. 6, respectively. Stress paths of sand matrix soils are shown in $p' - q$ diagrams. The critical state line (CSL) was defined by M which denotes the critical state parameter of the soils.

From Fig. 4, it can be seen that as the amount of fines content in coarse sand increases, the value of M decreases up to a certain amount of fines, then increased thereafter. M value decreases from M = 1.412 (at sample S100K0-C) to a minimal value of M = 1.348 (at sample S70K30-C), and after which increases to M = 1.395 (at sample S60K40-C). This finding is in agreement with the results of Marto et al. (2014) that also used clean sand and kaolin for reconstituted sand matrix soils. However the M values obtained was different as they used river sand. On the other hand, from the figures, increasing of fines leads to decreased of p' at the same relative density, Dr = 15% (loose state). A similar behaviour was observed by Simpson and Evans (2016) and Monkul et al. (2016).

Monotonic behaviour of sand matrix soils for medium and fine sands are shown in Figs. 5 and 6, respectively. Figure 5 demonstrates the monotonic behaviour of medium sand matrix soils, where the M achieved maximum value (M = 1.38) at 10% of fines contents, unlike the coarse sand matrix soils which reached the maximum value (M = 1.41) at 0% fine contents. M value was seen to decrease gradually thereafter with the increased of fines content without showing and threshold fines content value. Similar observation was also obtained for fine sand matrix soils, but the M reached maximum value at sample S80K20-F (20% fines content).

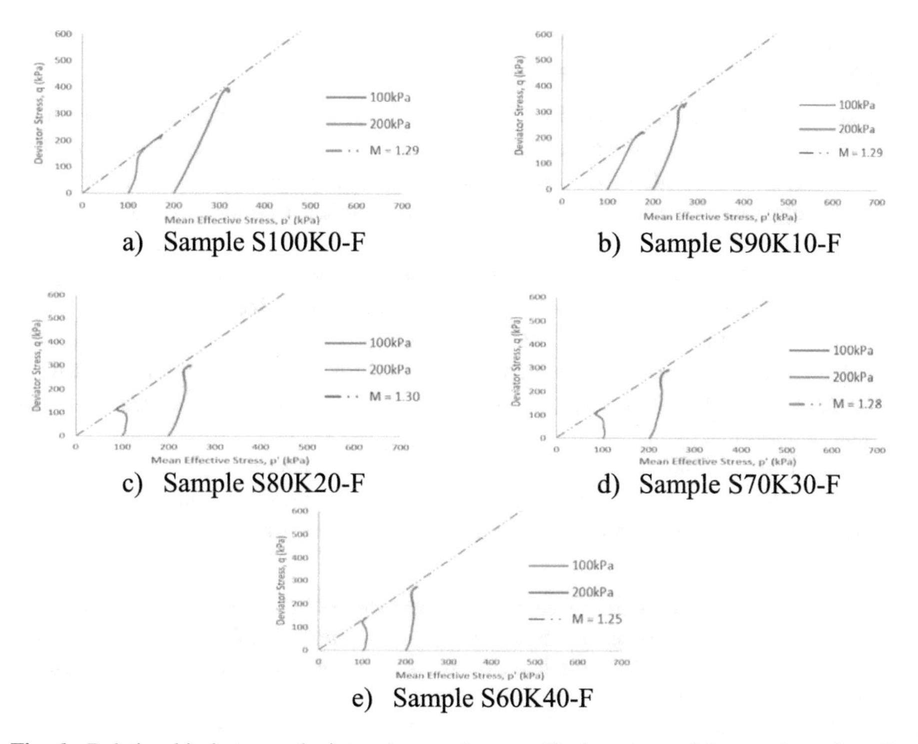

Fig. 6. Relationship between deviator stress and mean effective stress of fine sand matrix soils

Comparison were made between three different sizes of sand in terms of critical state parameter, M. It can be seen from Figs. 4, 5 and 6 that the M value of sand matrix soils was decreased with decreases of sizes of sand (coarse sand to fine sand). For example, sample S100K0-C with value of M = 1.41, and then decreased gradually for sample S100K0-M with value of M = 1.30, and also in sample S100K0-F with value of M = 1.29. In order to compare the effects of sand sizes on the sand matrix soils, the results reported by Tan (2015) that utilized Johor Sand with M ranges from 1.102 to 0.807, was studied. From the observation, the use of Johor (river) clean sand in sand matrix soils had resulted in lower M values compared to the M values obtained when different sizes of sand (coarse, medium, and fine) were used in this study. Hence, it shows the different behaviour of engineering properties of sand matrix soils with the application of various sizes of sand. This condition shows that grain sizes play an important role in explaining the behaviour of sand matrix soils.

4 Conclusion

This paper presents the findings on effects of sand sizes on the engineering behaviour of tropical sand matrix soils. The limitations of the study have been set at three sizes of sand, 0% to 40% of fines content, target relative density of 15% (loose state), effective consolidation stress at 100 kPa and 200 kPa, CSL shown in stress path

$(p' - q)$ diagram and undrained condition for monotonic triaxial test. The findings of this study are concluded as follows:

1. Basic properties results show that, as the fines content in sand matrix soils increases, the classification of soils changed from SP to SM after 20% of fines content. This was observed for all three sizes of the tropical sand in sand matrix soils. The maximum particle density was achieved at lower values of fines content for medium and fine sand matrix soils compared to coarse sand matrix soil. The sand size has significant effect on the maximum and minimum void ratio of sand matrix soils. At the same fines content, the values increased as the sand size decreased.
2. The critical state parameter, M value from monotonic test for coarse, medium, and fine sand matrix soils was measured at 1.41 to 1.35, 1.38 to 1.29, and 1.30 to 1.25, respectively. The maximum and minimum values of M differ between different sizes of sand.
3. The effects of different sizes of sand in sand matrix soils on engineering behaviour was found to be a good indicator as one of the significant contribution of soil behaviour to the liquefaction resistance in sand matrix soils.

Acknowledgements. The authors would like to acknowledge the Ministry of Higher Education Malaysia through the Fundamental Research Grant Scheme (FRGS) (No. PY/2016/07185, Vot. R.J130000.7822.4F849) in financing this research. Appreciation is expressed to the Faculty of Civil Engineering, Universiti Teknologi Malaysia for the facilities provided to undertake the experimental works. The second author also expresses his gratitude to the Malaysia-Japan International Institute of Technology (MJIIT), UTMKL for awarding Incentive Scheme to undertake his Ph.D. program.

References

Been, K., Jefferies, M.G., Hachey, J.: The critical state of sands. Geotechnique **41**(3), 365–381 (1991)

Been, K., Jefferies, M.: Stress–dilatancy in very loose sand. Can. Geotech. J. **41**(5), 972–989 (2004)

Benghalia, Y., Ali, Bouafia A., Canou, J., Dupla, J.: Liquefaction susceptibility study of sandy soils: effect of low plastic fines. Arab. J. Geosci. **8**, 605 (2015). https://doi.org/10.1007/s12517-013-1255-0

BS 1377: Methods of Test for Soils for Civil Engineering Purposes (1990)

Lade, P.V., Yamamuro, J.A.: Effects of nonplastic fines on static liquefaction of sands. Can. Geotech. J. **34**(6), 918–928 (1997). https://doi.org/10.1139/t97-052

Marto, A., Tan, C.S., Kasim, F., Mohd Yunus, N.Z.: Seismic impact in peninsular Malaysia. In: The 5th International Geotechnical Symposium-Incheon, Seoul, South Korea, pp. 237–242 (2013). https://doi.org/10.13140/2.1.3094.9129

Marto, A., Tan, C.S., Makhtar, A.M., Leong, T.K.: Critical state of sand matrix soils. Sci. World J. **2014**, 1–7 (2014). Article ID 290207. http://dx.doi.org/10.1155/2014/290207

Marto, A., Tan, C.S., Makhtar, A.M., Pakir, F., Chong, S.Y.: Effect of fines content on critical state parameters of sand matrix soils. In: AIP Conference Proceedings 1755, p. 060001 (2016). https://doi.org/10.1063/1.4958492

Monkul, M.M., Yamamuro, J.A.: Influence of silt size and content on liquefaction behavior of sands. Can. Geotech. J. **48**(6), 931–942 (2011)

Monkul, M.M., Etminan, E., Şenol, A.: Influence of coefficient of uniformity and base sand gradation on static liquefaction of loose sands with silt. Soil Dyn. Earthq. Eng. **89**(2016), 185–197 (2016)

Park, S., Kim, Y.: Liquefaction resistance of sands containing plastic fines with different plasticity. J. Geotech. Geoenviron. Eng. **139**(5) (2013). ©ASCE, ISSN 1090-0241/2013/5-825–830

Phan, V.T., Hsiao, D., Nguyen, P.T.: Effects of fines contents on engineering properties of sand-fines mixtures. Procedia Eng. **142**(2016), 213–220 (2016). https://doi.org/10.1016/j.proeng.2016.02.034

Rees, S.: What is triaxial testing? Part 1 of 3 (2013). Published on the GDS website www.gdsinstruments.com

Simpson, D.C., Evans, T.M.: Behavioral thresholds in mixtures of sand and kaolinite clay. J. Geotech. Geoenviron. Eng. (2016). © ASCE. https://doi.org/10.1061/(asce)gt.1943-5606.0001391

Tan, C.S., Marto, A., Leong, T.K., Teng, L.S.: The role of fines in liquefaction susceptibility of sand matrix soils. Electron. J. Geotech. Eng. **18L**, 2355–2368 (2013)

The Star News: Minister of Housing and Local Government Ministry (2012)

Wang, Y., Wang, Y.: Study of effects of fines content on liquefaction properties of sand. In: GeoShanghai 2010 International Conference. Soil Dynamics and Earthquake Engineering. ASCE (2010). https://doi.org/10.1061/41102(375)33

Prediction of Parallel Clay Cracks Using Neural Networks – A Feasibility Study

Tanveer Choudhury and Susanga Costa(✉)

School of Science, Engineering and IT, Federation University Australia,
Ballarat, VIC, Australia

Abstract. Cracking in drying clay soil is a common phenomenon especially in arid and semi-arid regions. Proper understanding and reliable prediction of the extent and nature of cracks in clay is vital for the design and construction of geo-infrastructures. While many models have been developed over the years to predict cracking, they are focused on a single crack rather than the whole network. This paper presents a feasibility study on a novel intelligent approach based on artificial neural network to predict the number of cracks in soil for a given combination of input parameters. Initial moisture content, specimen layer thickness and size of the specimen are used as inputs to the model. The output is the number of cracks. The collected database is used to train, validate and optimise the neural network models. The optimisation steps are discussed and analysed as the predicted number of cracks are compared to the experimental ones. A reasonable agreement was found between the experimental and predicted data. The results indicate that the model can be further improved to make more reliable predictions.

1 Introduction

Cracking in drying soil is a natural phenomenon, which is common in clay soils. When wet clay soil is exposed to high temperature and low humidity conditions, soil loses water. The removal of water from the soil matrix cause the soil to shrink. If the soil is not allowed to shrink freely due to the boundary conditions, as it happens almost everywhere in the field, tensile stresses are built inside the soil matrix, which eventually open up cracks. Tensile cracks in clay soil can be severely detrimental in many civil engineering applications such as earth dams, unsealed road pavements, landfill clay liners, mine tailings, cut slopes etc. (Peron et al. 2009).

The analysis of desiccation cracks in soil is not straightforward. Soil is a particulate material in contrast to many other continuum materials used in engineering constructions. While traditional theories such as elastic theory and linear elastic fracture mechanics can be utilised in analysis, their use is limited by vastly non-linear behaviour of soil. Almost all the properties of soil are highly sensitive to the amount of water present inside the soil matrix. Thus the constant removal of water, as it happens during drying, causes continuous changes in other properties making the analysis complex. Moreover, it has been established that the intensity and dimensions of cracks depend on the geometry of the soil mass and the drying rate. This further increases the complexity of the problem.

S. Hemeda and M. Bouassida (Eds.): GeoMEast 2018, SUCI, pp. 214–224, 2019.
https://doi.org/10.1007/978-3-030-01941-9_19

Over the last few decades, analysis, modelling and prediction of desiccation cracks in soil have gained the attention of researchers. Qualitative studies on soil cracking by several researchers provided the basis for understanding the mechanism of crack formation (Peron et al. 2009; Costa et al. 2013; Corte and Higashi 1960). Modelling and prediction of cracking involves the onset of cracking, intensity of cracking and depth and spacing between cracks. Notable attempts were made by past researchers (Morris et al. 1992; Konrad and Ayad 1997; Hueckel et al. 2014; Costa et al. 2018; Cost and Kodikara 2012; Amarasiri et al. 2011) to develop analytical and numerical models. While all these models took a step forward, they are still far from making reliable predictions under field conditions. This is largely due to extremely non-linear relationships between the parameters involved. Nevertheless, researchers continue to explore novel methods to develop more reliable models.

This paper investigate the feasibility of using intelligent prediction techniques using Artificial Neural Network (ANN) to predict soil cracking. ANN is a robust and adaptive predictive modelling technique that has been used to predict soil properties in the past: shear strength (Kiran et al. 2016; Chitra and Gupta 2014), compaction and permeability (Tizpa et al. 2015; Sinha and Wang 2008) and foundation settlement (Sivakugan et al. 1998). In this study, ANN is used to predict the number of cracks in a given specimen of soil under different geotechnical parameters that causes the phenomenon. This is achieved as the ANN is passed through training and optimisation process to establish input output parameter correlation through learning the underlying parameter relationships.

2 Simulation Model

2.1 Experimental Condition and Parameter Selection

The study was carried out using desiccation cracking experiments on thin, long soil specimens. The specimens used in these experiments have a depth to length ratio of 1:30 or higher. The width to length ratio also follows a similar value. Hence, the soil specimens are considered to be thin and long.

The advantage of using thin, long specimens is that they produce a simple crack pattern. Desiccation cracks on these specimens are predominantly parallel. More interestingly, the first crack can be expected at mid-length while the subsequent cracks appearing in a sub-dividing manner. Figure 1 shows the irregular crack patterns in a circular specimen and parallel patterns in a thin, long specimen. Several researchers have worked on thin, long soil specimens to study the desiccation cracks (Costa et al. 2008, 2018; Peron et al. 2013; Nahlawi and Kodikara 2006). It should also be noted that thin, long specimens are not completely hypothetical. There are field applications such as road pavements that illustrate behaviours similar to thing, long specimens.

There are several factors that can affect cracking in soil. Apart from environmental conditions such as temperature and relative humidity, soil type, specimen layer thickness and moisture content also affect the cracking process. In laboratory experiments, the geometry of the specimen also become important since the crack patterns are affected by the boundary conditions.

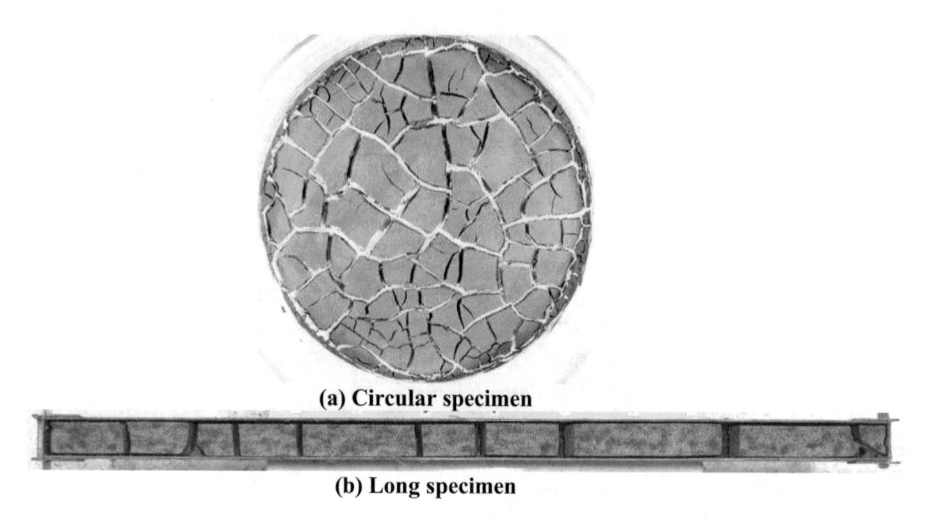

Fig. 1. Typical crack patterns on clay (a) circular specimen, (b) long specimen (Costa 2010).

Costa (2010) conducted systematic work on soil desiccation cracking which involved two types of tests: (i) free and (ii) restrained shrinkage. The fundamental relationship between the crack characteristics (e.g. number of cracks, depth, spacing and crack patterns) and various above mentioned influencing conditions were observed. From this study, the moisture content, drying rate and specimen thickness were identified as the primary controlling factors. In addition, the condition of interface between the soil and the mould at the base was found to be another significant factor affecting desiccation cracks.

Based on the literature (Peron et al. 2009, 2013; Corte and Higashi 1960; Hueckel et al. 2014), the input parameters to the ANN model were selected as: (i) initial moisture content (MC), (ii) soil layer thickness (LT), (iii) mould length (ML), and (iv) mould width (MW). The selected ANN output parameter was the number of cracks. It should be noted that in order to simplify the study, some input parameters were not considered. These include soil properties such as tensile strength, interface friction at base and drying conditions such as drying rate.

2.2 Database Collection

Database collection is an importance step in ANN modelling. A robust and sufficient large database is essential for the development of a well-trained network.

Database, DB, for this study was collected from the experimental study presented in Costa (2010). The experiments for desiccation cracking were conducted using Werribee clay, a material sourced from Werribee Victoria, Australia and consisted of 16 data points. Werribee clay has been classified as a highly reactive soil and its properties include LL = 127% and PL = 26%.

2.3 Database Processing

The database was normalized using Eq. (1) before passing through the neural network model. This linear transformation ensures that all process parameters are treated equally by the ANN and, thus, avoids calculation error relating to different parameter magnitudes.

$$X_{NORM} = \frac{X - X_{MIN}}{X_{MAX} - X_{MIN}} \quad (1)$$

In Eq. (1), X_{NORM} represents the normalized input parameter value and X the actual parameter value. X_{MAX} and X_{MIN} are the maximum and minimum possible values of the parameters based upon their physical limitations of the process. The physical limits of each input and output variable are given in Table 1. Table 1 further describes the physical limitation of the developed neural network model.

Table 1. Parameter physical limits

Variable	Lower limit	Higher limit
Initial moisture content (MC) (%)	31.6	134
Soil layer thickness (LT) (mm)	5	30
Mould length (ML) (mm)	250	600
Mould width (MW) (mm)	25	50
Number of cracks	4	11

2.4 Artificial Neural Network Model

The ANN is frequently implemented to model complex process relationships between the input and output parameters considering the variability, fluctuations and no prior assumptions. ANN consists of mathematical model of a group of interconnected artificial neurons and is used as a non-linear statistical data modelling tool.

ANN takes a connectionist approach to computation where the strength of each connections between the neurons is represented by the term 'weight' (Nelson and Illingworth 1991). These weights form the basis to process generalisation and evaluation of input and output parameter relationships. This can be achieved by proper optimisation of the weight matrix, which is achieved through a training process. The process of tuning weight matrix is called paradigm. The most powerful paradigm is the back-propagation paradigm, which is widely used (Fahlman 1988) and considered for this study.

The back-propagation paradigm used in this research is the Levenberg-Marquardt algorithm (Marquardt 1963). Standard back-propagation algorithms are: (i) very slow, (ii) require a lot of off-line trainings, and (iii) suffer from temporal instability as they tend to get stuck to the local minima (Adeloye and Munari 2006). Given a network not more than few hundred weights (which complies with the current model), the

Levenberg-Marquardt algorithm proves to be more efficient in comparison to the conjugate gradient and variable learning rate algorithm (Hagan and Mehnaj 1994).

To overcome the complex non-linear relationship between the output number of cracks and the selected input parameters, this research utilises a simple neural network model based on multi-layer perception (MLP) and back-propagation algorithm.

The ANN architecture consists of three main components: (i) input layer, (ii) output layer, and (iii) hidden layer. The input layer represents the aforementioned parameters associated with soil cracking – initial moisture content (MC), soil layer thickness (LT), mould length (ML) and mould width (MW). The output layer consist of the number of cracks.

The number of neurons, required in the ANN model to describe each physical parameter within the process is depended on the parameter nature. All the parameters under consideration in this study (MC, LT, ML, MW and number of cracks) are real valued parameters and are represented by one neuron each.

The hidden layer within ANN structure contributes most in establishing the process generalisation and parameter correlations. The number of hidden layer and number of neurons is each hidden layer is determined from the network optimisation process. The proposed ANN architecture is presented in Fig. 2.

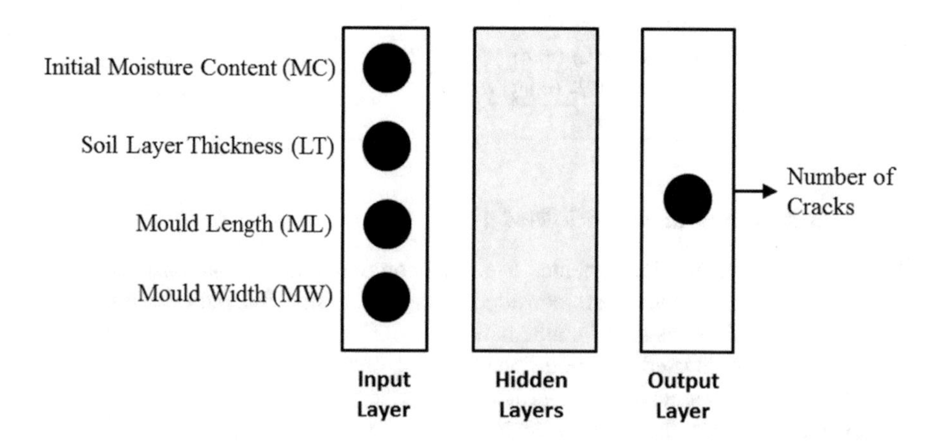

Fig. 2. The artificial neural network (ANN) architecture

2.5 Model Training and Optimization

All the ANN simulations in this study were performed with MATLAB (R2017a: The Math Works Inc., Natick, MA, USA). ANN training and optimisation is a cyclic process and uses the database, DB. The process decides on the number of hidden layers, number of neurons in each hidden layer in conjunction with the optimisation of the weight population to produce the lowest error performance function.

To achieve this objective, an initial weight population and the number of neurons in the hidden layers are assumed. The network response of the number of cracks is computed and compared with the actual values from DB to obtain the error performance function. Based on this value, the network weight matrix is re-computed and optimised with the aim to minimise the error function. This becomes part of the network training process, which is run for a number of iterations – epochs.

It is really important to set the optimal number of epoch and allow the model to train for sufficient amount of time so that it optimises the network and generalises the process underpinning the input output parameters. If the number of epochs is set too low, the network fails to learn the function, if set too high, the network tends to memorise the training data instead of generalising the function. The maximum number of epoch set in this study is 1000 with the log-sigmoid as the transfer function in all layers.

For the network training, optimisation and testing, DB is divided into three data sets: (i) 60% of the data allocated for network training (DB-TRN), (ii) 20% of the data allocated for network validation (DB-VAL), and (iii) 20% of the data allocated for network testing (DB-TST). For the test and validation step, it is important to note that the weight population and other network parameters are kept static as the network is being tested with the input data and checked for prediction performance.

The Mean Absolute Error (MAE) and correlation coefficient (R) are selected as the model's performance measuring functions. The MAE generated by the network on the test set provides a measure of the "generalisation error" of the trained network models. The lower this error value, better is the network's ability to generalise process and predict infiltration rate with sufficient accuracy under unknown conditions.

The correlation coefficient, R, values on the test set provides an understanding of how well the trained network's response, to the unseen inputs, fits with the respective experimental infiltration rate value. Larger R values represents better fit and model performance in generalising the process dynamics.

The performance of a trained network is sensitive to the size of the hidden layers. To overcome the extreme non-linearity associated with the geotechnical process considered in this study, six hidden layers were used. For the initial model, the number of hidden layer neurons were set to six, five, four, three, two and one in each of the hidden layers starting from one to six (6-5-4-3-2-1). For the subsequent models, the number of neurons in each hidden layer was increase by one and simulations were carried out for different ANN models. The maximum hidden layer neurons used was 25-24-23-22-21-20.

For each of the developed models, the networks were trained with the training set (DB-TRN) to find the optimised weight matrix, validated with the validation set (DB-VAL) to ensure there is no over-fitting (Choudhury et al. 2012) and tested with the test set (DB-TST) to check for network's generalisation performance. The network trainings were repeated ten times, in each case, and the network generating minimum MAE value on DB-TRN was stored and saved.

3 Results and Discussion

The variations of R-values of all the trained ANN models with six hidden layers on DB-TRN is presented in Fig. 3. The average R-value for all the networks was found to be 0.6942 with standard deviation of 0.0785. Observing Fig. 3 and the standard deviation, it can be seen that the performance of different trained networks has been fluctuating. However, considering all the developed ANN models, the network with 15-14-13-12-11-10 hidden layer neurons was found to generate the maximum R-value of 0.8817.

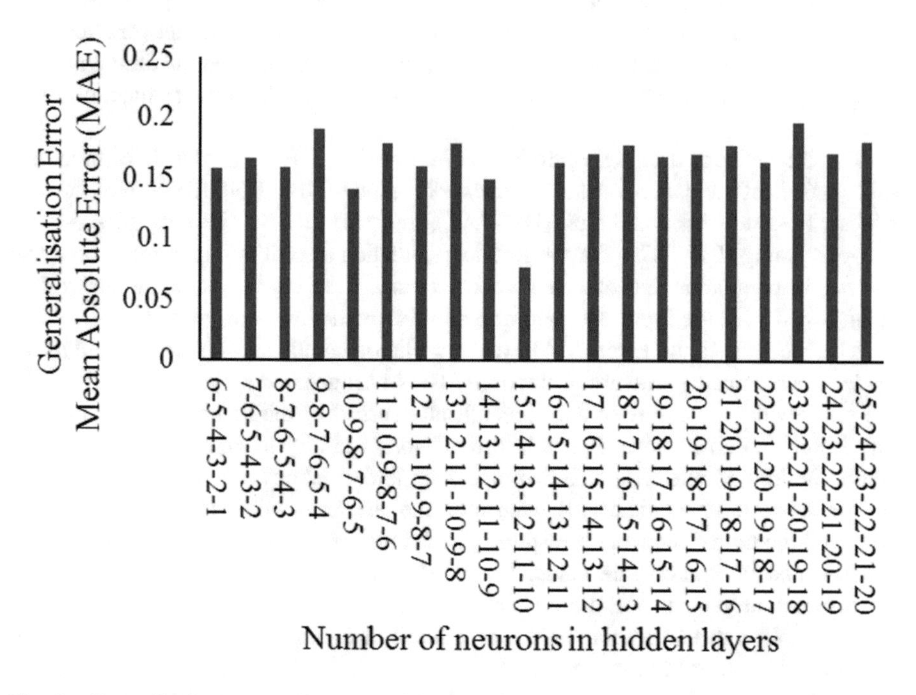

Fig. 3. Generalisation error variations of different ANN models on the training taset (DB-TRN).

The generalization error for different developed ANN models (Fig. 4) showed similar trend in fluctuations of performance values shown in Fig. 3. The average MAE value was 0.1653 with standard deviation of 0.0239. The minimum generalisation error of 0.0772 was obtained for the network with 15-14-13-12-11-10 hidden layer neurons. This is in harmony with results shown in Fig. 3.

The R-value variations for developed ANN models on DB-TST are presented in Fig. 5. The maximum R-value was 0.9998 for the network with 15-14-13-12-11-10 hidden layer neurons. The average R-value was computed to be 0.9361 with standard deviation of 0.0917. The performance of the developed ANN models, in terms of the R-value, was found to be better when tested with DB-TST. The network with 15-14-13-12-11-10 hidden layer neurons generated maximum R value which 13% higher on DB-TST (Fig. 5) than that when trained with DB-TRN (DB-TRN). This clearly indicates the strong generalisation ability of the developed ANN model.

The minimum generalisation error (MAE) of all the developed ANN models on DB-TST was 0.1114. The variations of these values are presented in Fig. 6. Compared to the generalisation error performance of the developed models on DB-TRN,

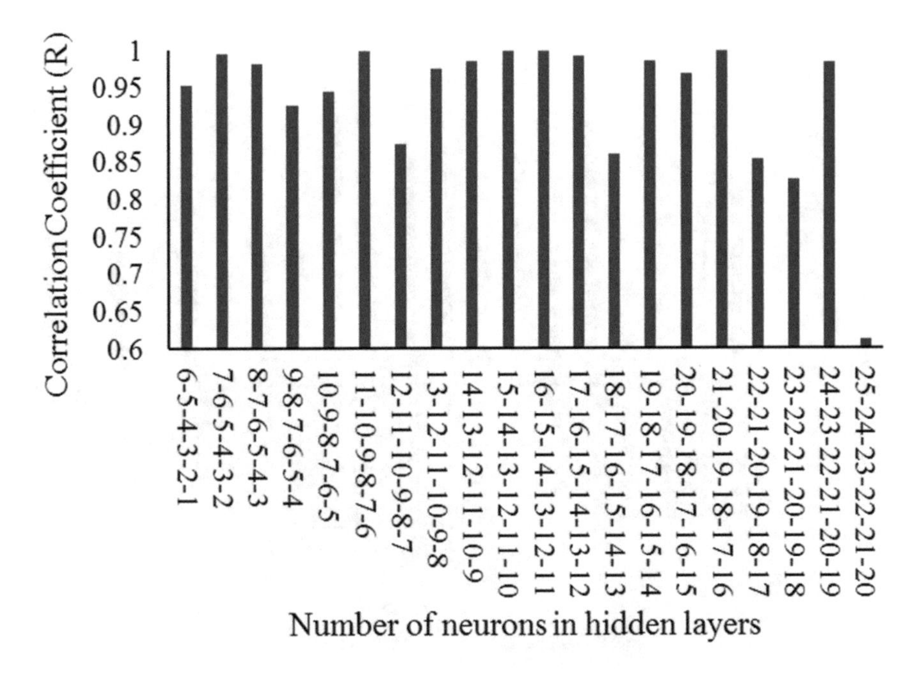

Fig. 4. Correlation coefficient (R) variations of different ANN models on the test dataset (DB-TST).

the average generalisation error for all the networks on DB-TST was slightly higher value of 0.2532 in comparison to the value of 0.1653 for the earlier. Comparing all the network performance on DB-TRN and DB-TST, the inferior MAE value and the superior R-value indicates the development of ANN models who has correctly learned the underlying input-output parameter relationships and have been able to predict correct trend in output variables with the corresponding fluctuations to input values. However, there is certain degree in under and/or over prediction by the models.

Considering all the performance, the network with 15-14-13-12-11-10 is referenced as “NET” to be used for further analysis.

NET was used to simulate the entire dataset (DB) to predict the number of cracks under different input conditions. A plot of the predicted values along with the experimental (actual) values from DB is presented in Fig. 7. It can be seen that the predicted number of cracks are in reasonable agreement with the experimental results. Number of cracks is helpful to get an indication of the nature of the crack pattern that can be expected in a given soil mass under specified drying conditions. However, in order to understand the total effect of a crack network in a soil mass, three dimensional aspects of cracks should be considered. For this, a statistical parameter such as the average uncracked area or ratio between the cracked and uncracked volumes need to be used as the output parameter.

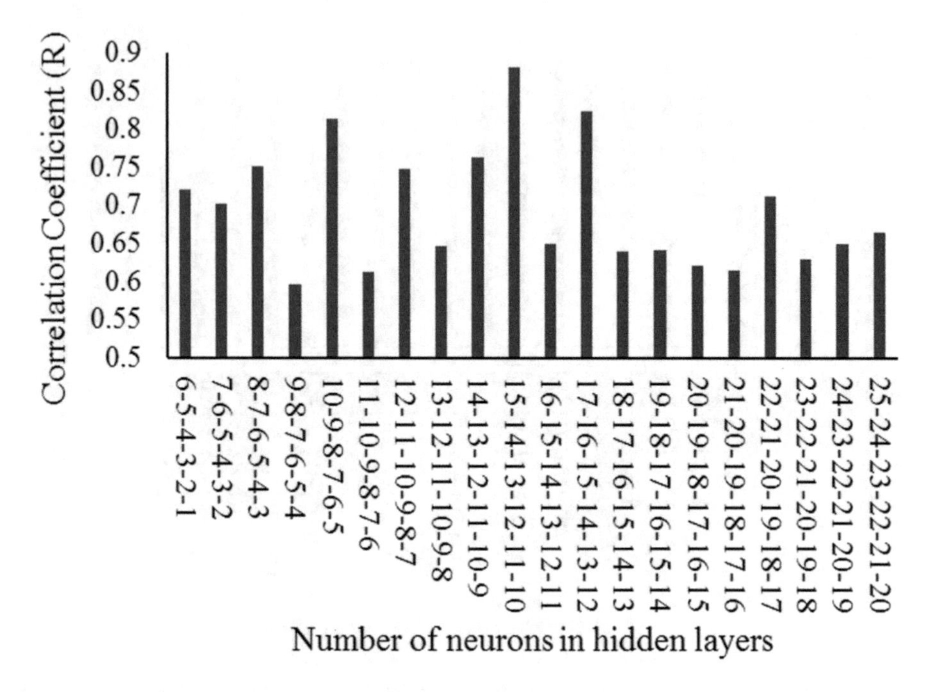

Fig. 5. Correlation coefficient (R) variations of differnet ANN models on the training dataset (DB-TRN).

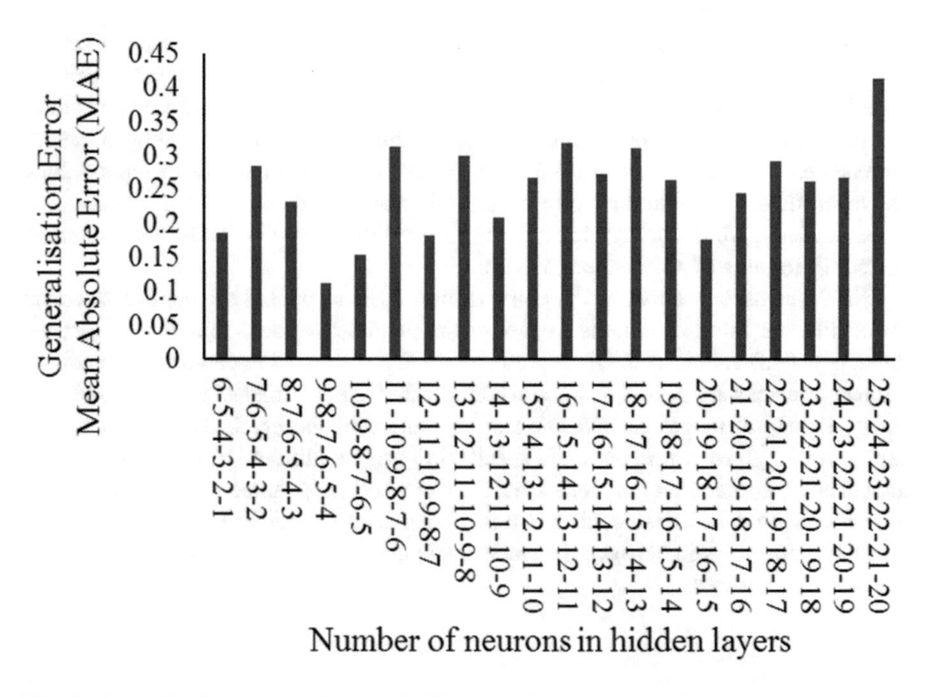

Fig. 6. Generalisation error variations of different ANN models on the test dataset (DB-TST).

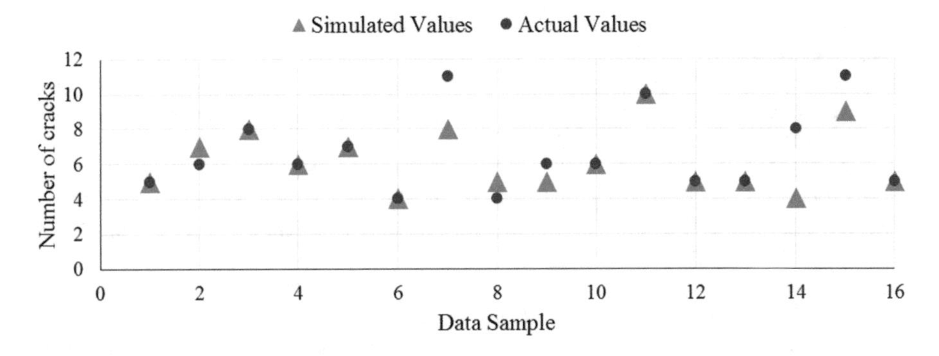

Fig. 7. Prediction of the selected ANN model NET on the total dataset (DB)

4 Conclusions

Artificial neural network was used to model the highly non-linear problem of desiccation crack formation in clay soils. A relatively simple case of clay cracking, namely parallel cracks occurring in thin, long clay layers, was modelled. The database consisted of laboratory experimental results of desiccation cracking tests. Four key parameters that affect clay crack formation were used as input parameters.

The developed Artificial Neural Network (ANN) model predicted the number of cracks in a specimen. The ANN model was based on Multi-Layer Perception (MLP) architecture and was trained with Levenberg-Marquardt back-propagation algorithm. To learn the input-output parameter relationship and handle the extreme non-linearity of the geotechnical process considered in this study, six hidden layers were selected. Analysing both training and testing performance, the network with 15-14-13-12-11-10 hidden layer neurons were selected as the best performing network which was used to simulate the entire collected database (DB) to predict the number of cracks. The performance was very encouraging with close agreement was evident between the test results and predicted data. Results further provide applicability of ANN technique in predicting cracks in soils.

Future research will focus on different types of cracks including irregular cracks occurring under field conditions. More parameters, such as properties of clay, need to be incorporated as input parameters. A statistical parameter that can describe both the horizontal and vertical extent of cracks, such as crack intensity factor, will be a more accurate output parameter.

References

Adeloye, A.J., Munari, A.D.: Artificial neural network based generalized storage-yield-reliability models using the Levenberg-Marquardt algorithm. J. Hydrol. **326**, 215–230 (2006)

Amarasiri, A.L., et al.: Numerical modelling of desiccation cracking. Int. J. Numer. Anal. Meth. Geomech. **35**, 82–96 (2011)

Chitra, R., Gupta, M.: Neural networks for assessing shear strength of soils. Int. J. Recent. Dev. Eng. Technol. **3**, 24–32 (2014)
Choudhury, T.A., et al.: Improving the generalization ability of an artificial neural network in predicting in-flight particle characteristics of an atmospheric plasma spray process. J. Therm. Spray Technol. **21**, 935–949 (2012)
Corte, A., Higashi, A.: Experimental research on desiccation cracks in soil-research Report 66. In: Wilmette, I. (ed.) US Army Snow Ice and Permafrost Research Establishment (1960)
Costa, S., Kodikara, J.: Evaluation of J integral for clay soils using a new ring test. Geotech. Test. J. **35**, 981–989 (2012)
Costa, S.: Study of desiccation cracking and fracture properties of clay soils. Ph.D. Dissertation, Department of Civil Engineering, Monash University (2010)
Costa, S., et al.: Modelling of desiccation crack development in clay soils. In: Proceedings of the 12th International Conference of IACMAG, Goa, India, pp. 1099–1107 (2008)
Costa, S., et al.: Salient factors controlling desiccation cracking of clay in laboratory experiments. Geotechnique **63**, 18 (2013)
Costa, S., et al.: Theoretical analysis of desiccation crack spacing of a thin, long soil layer. Acta Geotech. **13**(1), 1–11 (2018)
Fahlman, S.E.: Faster-learning variations on back propagation: an emperical study. In: Proceedings of the 1988 Connectionist Models Summer School, pp. 38–51 (1988)
Hagan, M.T., Mehnaj, M.B.: Training feedforward networks with the Marquardt algorithm. IEEE Trans. Neural Netw. **5**, 989–993 (1994)
Hueckel, T., et al.: A three-scale cracking criterion for drying soils. Acta Geophys. **62**, 1049–1059 (2014)
Konrad, J.-M., Ayad, R.: A idealized framework for the analysis of cohesive soils undergoing desiccation. Can. Geotech. J. **34**, 477–488 (1997)
Marquardt, D.: An algorithm for least-squares estimation of nonlinear parameters. J. Soc. Ind. Appl. Math. **11**, 431–441 (1963)
Morris, P.H., et al.: Cracking in drying soils. Can. Geotech. J. **29**, 263–277 (1992)
Nahlawi, H., Kodikara, J.: Laboratory experiments on desiccation cracking of thin soil layers. Geotech. Geol. Eng. **24**, 1641–1664 (2006)
Nelson, M.M., Illingworth, W.T.: A Practical Guide to Neural Nets. Addison-Wesley Publishing Company Inc., USA (1991)
Peron, H., et al.: Fundamentals of desiccation cracking of fine-grained soils: experimental characterisation and mechanisms identification. Can. Geotech. J. **46**, 1177–1201 (2009)
Péron, H., et al.: Formation of drying crack patterns in soils: a deterministic approach. Acta Geotech. **8**, 215–221 (2013)
Sinha, S.K., Wang, M.C.: Artificial neural network prediction models for soil compaction and permeability. Geotech. Geol. Eng. **26**, 47–64 (2008)
Sivakugan, N., et al.: Settlement predictions using neural networks. Aust. Civ. Eng. Trans. **40**, 49 (1998)
Tizpa, P., et al.: ANN prediction of some geotechnical properties of soil from their index parameters. Arab. J. Geosci. **8**, 2911–2920 (2015)

Compaction Properties of Cement Kiln Dust

Mahmoud E. Hassan[1(✉)], Ayman L. Fayed[2], and Mohamed Y. Abd El-Latif[2]

[1] Construction and Building, 6th October University, Giza, Egypt
mahmoud_elsayed_151189@hotmail.Com
[2] Geotechnical Engineering Dept, Ain Shams University, Cairo, Egypt

Abstract. Cement kiln dust (CKD) is a by-product of cement factories representing a real challenge for treatment due to the huge produced quantities of severe environmental hazards. Disposing of CKD requires compacting the raw material to the minimum volume due to the high cost of landfills. As the compaction of such material is not a common process especially in Egypt, an extensive experimental study program was planned and executed in order to define the best way of compacting the CKD.

Different factors affecting the compaction process were investigated in the current experimental study including the type of the mixing fluid and the effect of pre-curing of the CKD before compacting it either by fully submergence in water for a certain period, or by wetting it with a certain amount of water, and allowing certain periods for the chemical reaction to take place prior to drying and performing the compaction process.

Results of the performed experimental study showed that, using of salt water/seawater provides the optimum way for compacting the CKD to the maximum possible dry density at reasonable optimum water content. Using of waste used engines oil resulted in better efficiency of the compaction process in terms of the achieved maximum dry density with very limited amount of oil of about 0.205%, however oil is not preferred as a mixing fluid due its environmental hazards and the anticipated high cost as well.

Pre-curing/wetting of the CKD with water prior to compaction didn't show a benefit, while on the contrary, entails using larger amount of water in the compaction process while achieving less maximum dry density compared to the normal case without pre-curing.

Due to the proximity of most of the cement factories in Egypt to the sea, the experimental study confirmed that using the salt water in compacting the CKD is the most favorable method. Also, the tests proved the ability to compact the CKD and hence disposal in landfills of reasonable volumes.

1 Introduction

Cement manufacturing is a critically important industry in the world. Egypt is one of the largest cement producers all over the world and the leading country in the Middle East, Africa, and the Arabian Region with a total production capacity of 70 million tons of clinker annually. Cement bypass dust (CBPD), also known as cement kiln dust (CKD), is a fine cementitious powder that is produced in large amounts as a by-product

S. Hemeda and M. Bouassida (Eds.): GeoMEast 2018, SUCI, pp. 225–239, 2019.
https://doi.org/10.1007/978-3-030-01941-9_20

during cement manufacturing. CBPD is mainly composed of oxidized, anhydrous, micron-sized particles, and is considered as a major health hazard and hence recommended to be utilized in beneficial uses or to be safely disposed. Many research works are conducted all over the world, to find economical and efficient ways for using the CBPD in various applications like; soil stabilization, pavements, landfills and concrete mixes. The main challenge however, is that not all the produced amounts of the CBPD are able to be recycled, where in the United States only, for example, more than four million tons are yet required to be disposed annually, (Todres et al. 1992). Currently, one of the Egyptian cement factories located at Ain Sokhna has intended, for the first time in Egypt, to dig landfills to bury this dust and minimizes its environmental hazards. Burial of CBPD necessitate compacting it, by special ways and procedures, to decrease its volume to the minimum, in order to contain the high produced amounts of dust in reasonable size landfills. In this research work, different possible ways of compacting the CBPD are investigated, as a new technique of dealing with such product in Egypt, to dispose the large amount of dust in safe and environmentally friendly manner.

1.1 Materials

Different materials have been used during the scope of this work including the following:

1.2 Cement Kiln Dust

The sample of cement kiln dust used in this research was obtained from (ALARABIA FOR CEMENT, in Ain Sokhna zone) factory on 26th July, 2017. The used CKD was generally in a fresh condition. Figure 1 shows the form of the utilized CKD sample.

Fig. 1. Cement kiln dust sample (CKD) form

1.3 Fresh Water

Fresh water utilized in the experimental work is the ordinary tap water.

1.4 Salt Water

Used salt water consisted of fresh tap water mixed with 35 parts per thousand of normal table salt, composed primarily of sodium chloride (NaCl). The used salt water is basically a simulation of the normal seawater with almost the same salinity.

1.5 Waste Oil

Used cars' engine oil or lubricant fluids, after certain duration of functioning, is usually disposed due to the change in its viscosity and other physical and mechanical properties. Such waste oil is classified as a very harmful and environmentally hazardous. Many studies worldwide tried to treat or recycle this material in a safe manner to be reused again. In this paper, waste oil is investigated to be used in compacting the CKD.

2 Experimental Program

Details of the planned and executed experimental program are shown in Table 1 including three groups of tests. Initially, chemical and physical tests for the CKD samples such as the chemical analysis, grain size distribution and specific gravity tests were performed for characterization of the used material. After the material characterization, three groups of tests were conducted to investigate the effect of the examined different factors on the compaction process of the CKD. The investigated different factors in relation to the compaction characteristics are; the type of liquid used in mixing the CKD, the duration of full submergence of CKD samples in fresh water prior to compaction and the chemical reaction duration (curing duration) of CKD mixed with certain amount of water content. All tests were performed according to the standards mentioned in Table 2.

3 Test Results

3.1 Cement Kiln Dust Characterization Tests

Chemical Composition

Chemical analysis test was conducted in the Arabian Cement factory in Ain sokhna – Suez, using X-rays test by the winxrf equipment. The chemical composition was as given in Table 3 for the CKD sample by weight. The amount of the different chemical elements in the CKD can vary significantly from one plant to another depending on the raw materials and type of collection process (Miller et al. 2003).

Table 1. Experimental tests program

Group no.	Test no.	Studied factor	Mixing liquid type	Test equipment	Test duration	Sample condition and composition
G-1	1–1	Mixing liquid type effect on the maximum dry density	Fresh water	Standard Proctor apparatus	Following the standards requirements	Dry CKD mixed with fresh water
	1–2		Salt water			Dry CKD mixed with salt water
	1–3		Oil			Dry CKD mixed with used oil
G-2	2–1	Effect of the duration of soaking CKD in water on the maximum dry density	Fresh water	Standard Proctor apparatus	Soaking for 1day	Fully submerged CKD in fresh water
	2–2				Soaking for 7days	
	2–3				Soaking for 30day	
G-3	3–1	Effect of the pre-wetting duration for chemical reaction on the maximum dry density	Fresh water	Standard Proctor apparatus	1 h of pre-wetting for chemical reaction	Dry CKD mixed with 10% fresh water
	3–2				12 h of pre-wetting for chemical reaction	
	3–3				24 h of pre-wetting for chemical reaction	

Table 2. Standards of the performed tests on the CKD samples

Type of test	CKD
Chemical composition	ASTM C114
Particle size distribution	ASTM D422
Specific gravity	ASTM D-854-92
Compaction test	Standard proctor method AASHTO T-99

Particle Size Distribution

Particles size distribution test was conducted on the CKD samples using the (151-H) hydrometer in accordance with the standards shown in Table 2. The grain size distribution curve results for CKD sample is shown in Fig. 2 representing a generally fine material with almost 90% in the size of silt (0.002 > D > 0.06).

Table 3. Chemical composition of the tested CKD samples

SiO_2	Al_2O_3	Fe_2O_3	CaO	MgO	SO_3	K_2O	Na_2O	Cl	LOI
13.857	4.528	2.986	56.909	2.966	2.18	5.515	1.488	7.459	2.112

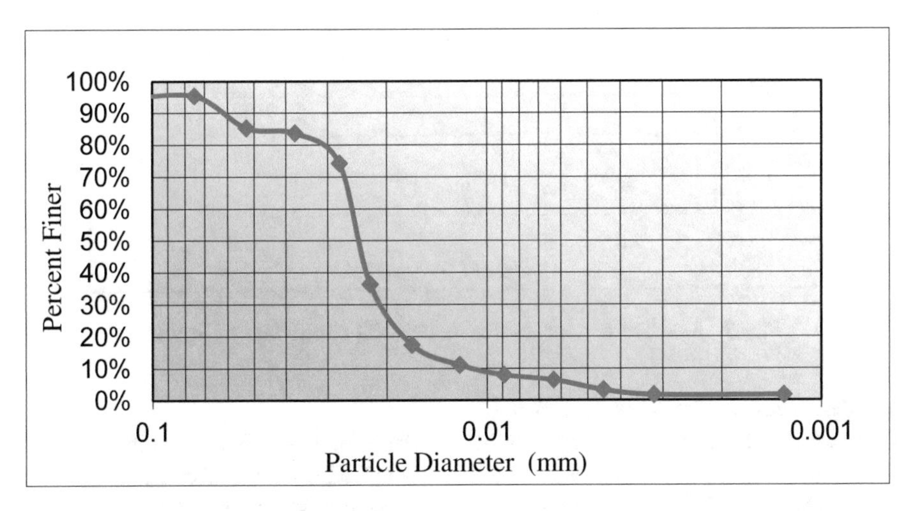

Fig. 2. Grain size distribution curve for the CKD sample

Specific Gravity Test (Gs)

The test was conducted on three CKD samples giving an average Gs value of approximately 2.738. The specific gravity of CKD is typically in the range of 2.6–2.8, (Baghdadi et al. 1995), which is usually less than that of the Portland cement (Gs $\sim$ 3.15).

3.2 Compaction Tests on the Cement Kiln Dust (CKD)

Compaction tests were conducted on the CKD samples to investigate the effect of several factors on the compaction efficiency and performance. The investigated factors included the type of liquid mixed with the CKD as a lubricant agent, the soaking duration of the CKD samples in water, and the chemical reaction period of CKD samples wetted with certain amount of water. Standard proctor test following the AASHTO T-99 requirements was adopted in all tests.

Standard Proctor Test with Fresh Water

Figure 3 shows the relation between dry density and water content (compaction curve) for the tested CKD sample blended with fresh water as a mixing liquid. The obtained maximum dry density and corresponding optimum water content resulting from the test are 1.319 gm/cm^3 and 28.979% respectively.

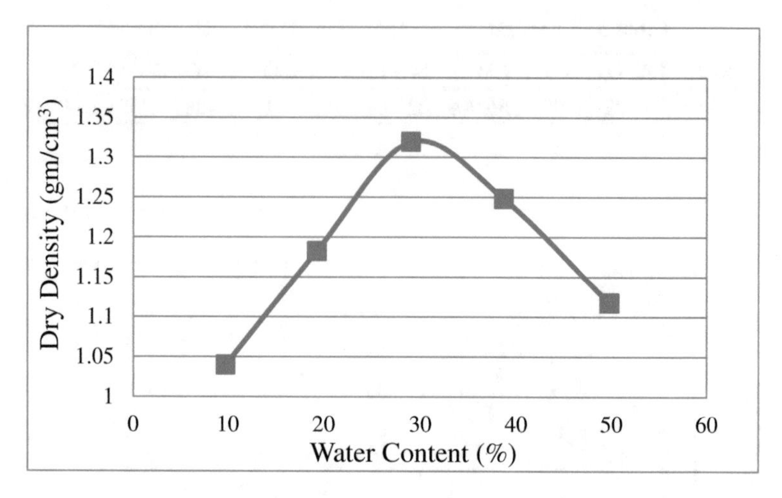

Fig. 3. Standard proctor test curve for CKD mixed with fresh water

Standard Proctor Test with Salt Water

To investigate the effect of using salt water, similar to seawater, instead of the costly tap fresh water on the compaction characteristics of the CKD, salt water consisting of fresh water mixed with 35 parts per thousand of NaCl similar to the sea water salinity was used. Results of the standard proctor test performed using the salt water are shown in Fig. 4. The obtained maximum dry density and corresponding optimum moisture content for this case are 1.342 gm/cm^3 and 30.323% respectively.

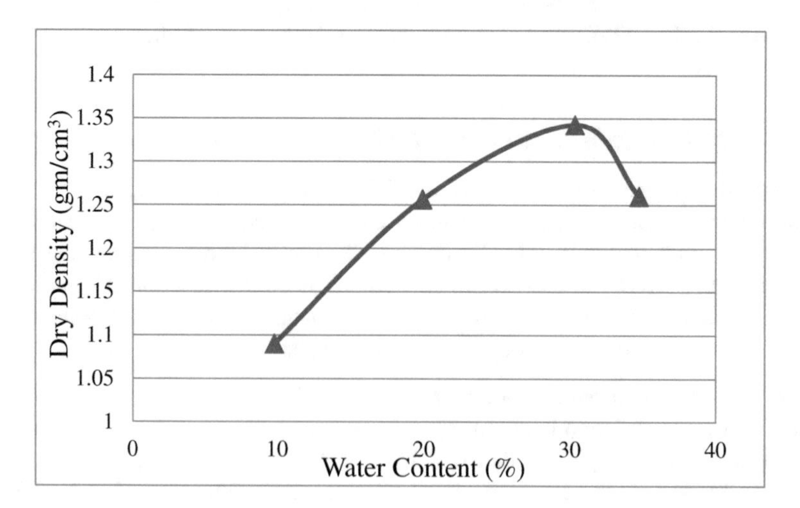

Fig. 4. Standard proctor test curve for CKD mixed with salt water

Standard Proctor Test with Waste Oil

The third investigated mixing fluid type in the compaction process of the CKD was the waste/used oil of cars' engines. Figure 5 shows the dry density- oil content relation. The achieved maximum dry density and corresponding optimum oil content resulting from the tests are 1.491 gm/cm^3 and 0.205% respectively.

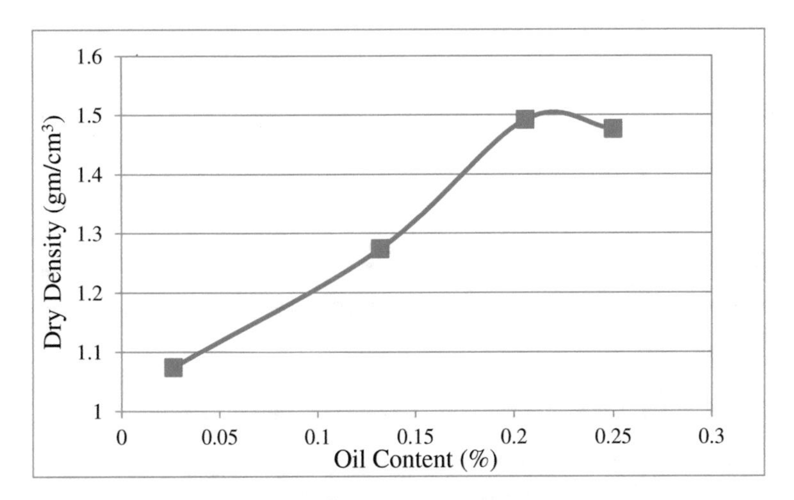

Fig. 5. Standard proctor test curve for CKD mixed with waste oil

Effect of the Full Submergence (Soaking) Duration of CKD in the Mixing Fluid on the Compaction Characteristics

In order to investigate the effect of the submergence duration of the CKD in the mixing fluid prior to starting the compaction process, fresh water was only used during all tests while the immersion duration was changed. CKD samples were immersed in fresh water for 1 day, 7 days and 30 days and then were dried in the oven for 24 h at 105 °C. After drying, the samples were grinded and then tested in the standard proctor apparatus with the same procedures adopted previously. Figure 6 shows the compaction curves for the three duration of submergence. From Fig. 6 and Table 4, summarizing the results, the recorded maximum dry densities and corresponding optimum water contents after 1 day, 7 days and 30 days are (0.988 gm/cm^3 and 51.506%), (0.93 gm/cm^3 and 49.281%) and (0.876 gm/cm^3 and 48.66%) respectively.

Effect of the Chemical Reaction Curing Period of CKD with Certain Fresh Water Content on the Compaction Characteristics

The effect of curing time duration after mixing the CKD samples with certain amount of fresh water and before commencing the compaction process is studied in this paper. Samples of the cement kiln dust were mixed with 10% by weight of fresh water and left for chemical reaction and curing for three different durations of 1, 10 and 24 h. After that, the samples were dried in the oven for 24 h at 105 °C and then subjected to the standard proctor test with the same previously adopted procedures. Figure 7 shows the compaction curves resulted from the standard proctor tests for the three different curing

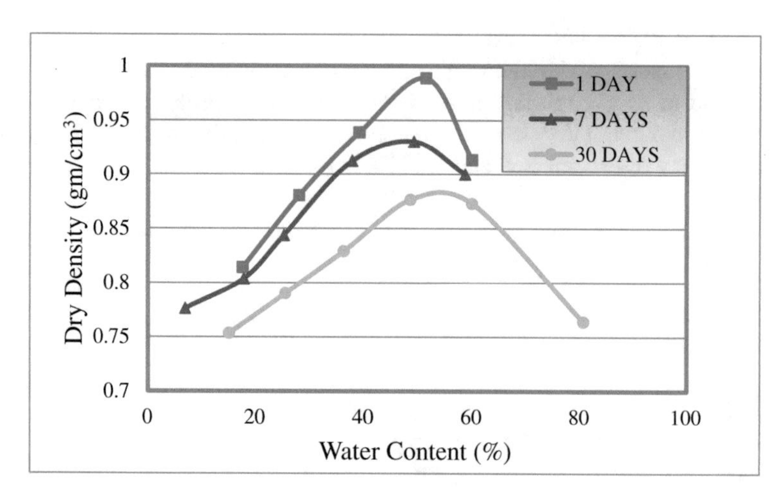

Fig. 6. Compaction curves after submergence of CKD samples in fresh water for 1 day, 7 days and 30 days

Table 4. Summary of the performed standard proctor compaction test results

Condition of CKD sample before performing the compaction test	Standard proctor test					
	Fresh water		Salt water		Oil	
	$\gamma_{d\ (max)}$, (gm/cm^3)	$w_{c\ (opt.)}$, (%)	$\gamma_{d\ (max)}$, (gm/cm^3)	$w_{c\ (opt.)}$, (%)	$\gamma_{d\ (max)}$, (gm/cm^3)	$w_{c\ (opt.)}$, (%)
Dry	1.319	28.979	1.342	30.323	1.491	0.205
Soaked in fresh water for 1 day	0.988	51.506	–	–	–	
Soaked in fresh water for 7 days	0.93	49.281	–	–	–	–
Soaked in fresh water for 30 days	0.876	48.660	–	–	–	–
Mixed with 10% water content and left for 1 h prior to drying and compaction	1.256	35.089	–	–	–	–
Mixed with 10% water content and left for 10 h prior to drying and compaction	1.136	42.415	–	–	–	–
Mixed with 10% water content and left for 24 h prior to drying and compaction	1.112	44.616	–	–	–	–

duration investigated. Summary of the tests results is presented in Table 4. The achieved maximum dry density and corresponding optimum water content for curing periods of 1 h, 10 h and 24 h are (1.256 gm/cm^3 and 35.089%), (1.136 gm/cm^3 and 42.415%) and (1.112 gm/cm^3 and 44.616%) respectively.

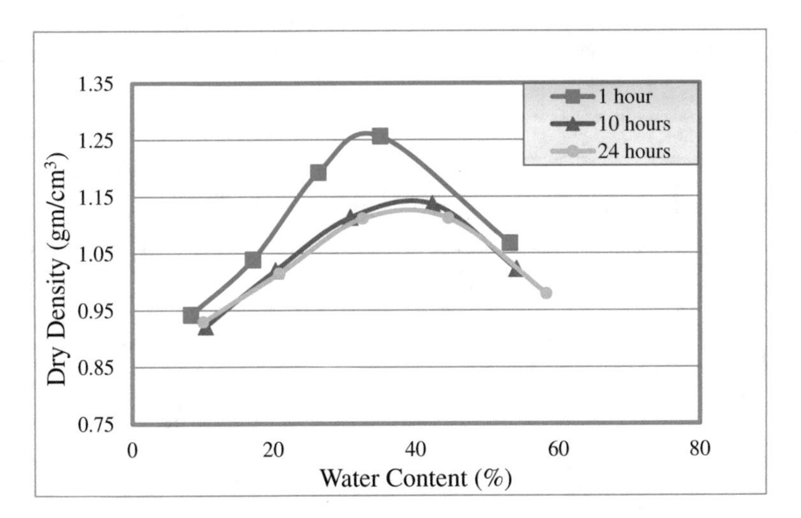

Fig. 7. Compaction curves for CKD samples mixed with 10% Wc and left for 1, 10 and 24 h. for chemical reaction

4 Comparisons and Discussions

Results of the performed compaction tests, under different conditions, are discussed in the following sections in order to understand the effect of the different studied factors on the compaction characteristics of the CKD. The cement kiln dust material showed different behaviors in the standard proctor compaction test while changing either of the type of mixing liquid, duration of submergence in fresh water, and duration of pre wetting with certain amount of fresh water prior to conducting the compaction test.

4.1 Effect of the Type of Mixing Fluid on the Compaction Test Results and Characteristics

Figure 8 shows the three compaction curves for the CKD samples tested using fresh water, salt water and waste oil. Table 4 shows the summary of the tests results regarding the achieved maximum dry density and optimum fluid content. From the obtained results, the achieved maximum dry density, $\gamma_{d\ max.}$, (was 1.319 gm/cm^3 and the corresponding optimum water content, $w_{c(opt.)}$, was 28.979%, when using the fresh water. For the case of mixing with salt water, the obtained maximum dry density increased to become 1.342 gm/cm^3 and with a corresponding $w_{c(opt.)}$, that was increased as well to become 30.323%. When using the waste oil as a mixing liquid, the max dry density increased further to become 1.491 gm/cm^3, while the recorded optimum fluid content (Oc) was as low as 0.205%. It could be concluded then that using salt water instead of fresh water led to an increase in the maximum dry density by about 1.74% while increasing the optimum water content by about 4.64%. This means that, when the cement plant is near to the sea with the availability of getting the seawater at a lower cost than the fresh water, the seawater can be then used in compacting the CKD efficiently to a maximum dry density that is almost equivalent to, or even higher than, that achieved using the valuable fresh water, while at less cost.

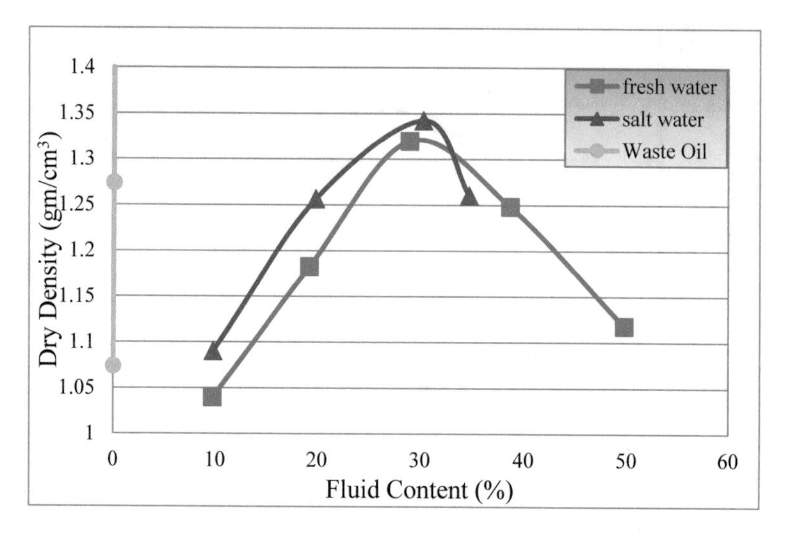

Fig. 8. Effect of the type of mixing fluid on the compaction test characteristics

From the results as well, using the waste oil in compacting the CKD showed very promising results, as while using a very low oil content that is about 0.205%, the achieved maximum dry density was about 13.04% and 11.10% higher than those corresponding to the fresh water and salt water conditions respectively. This means that using the waste oil in compacting the CKD will result in achieving two goals;

(1) Disposing of two waste materials in one operation, (2) compacting the CKD to higher density which means less storing volume compared to using either the fresh water or salt water. However, the cost of getting the waste oil to the cement plant should be considered in the financial study. Also, while the positive factor is the very limited amount of oil needed for compacting the CKD, the environmental hazards of using the oil in this process should be considered. However, using insulated landfills preventing the oil from penetrating the deep layers of the ground and reaching the groundwater could safely mitigate any expected environmental hazards.

4.2 Effect of the Full Submergence (Soaking) Duration on the Compaction Test Results and Characteristics

Figures 9 and 10 shows the variation of both the achieved maximum dry densities and corresponding optimum water contents of the standard proctor tests performed on CKD samples soaked prior to drying and compaction in fresh water for 1 day, 7 days and 30 days. Summary of the results are given in Table 4. It can be noticed that, generally, increasing the soaking time results in decreasing both the maximum dry density and corresponding $w_{c(opt.)}$. However, by comparing between the case of compaction of the CKD using fresh water without soaking and the case of soaking for different durations, it can be seen that, soaking the samples for 1 day resulted in decreasing the achieved maximum dry density by about 25.09% while increasing the $w_{c(opt.)}$ by about 77.74%. For the case of soaking for seven days, the achieved maximum dry density decreased by 29.49% while increasing the $w_{c(opt.)}$ by 70.06%. For the case of soaking for thirty days, the achieved maximum dry density decreased by 33.59% while increasing the $w_{c(opt.)}$ by 67.91%.

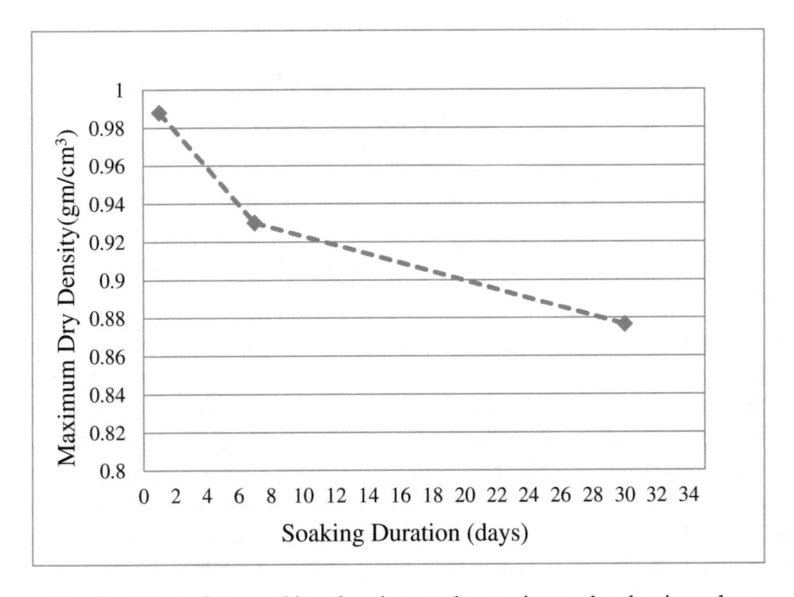

Fig. 9. Effect of the soaking duration on the maximum dry density value

From these results, the maximum dry density is constantly decreasing by increasing the soaking period and the optimum water content is generally increased by soaking, but the degree of increasing is declining by rising the soaking duration. Accordingly, it is not recommended to soak the CKD in water prior to compaction as the achieved density after compaction will be decreasing which means higher volume and storage space will be required. Furthermore, soaking resulted in increasing the amount of water needed for compaction (optimum water content), in addition to the large amount of water needed for the soaking process.

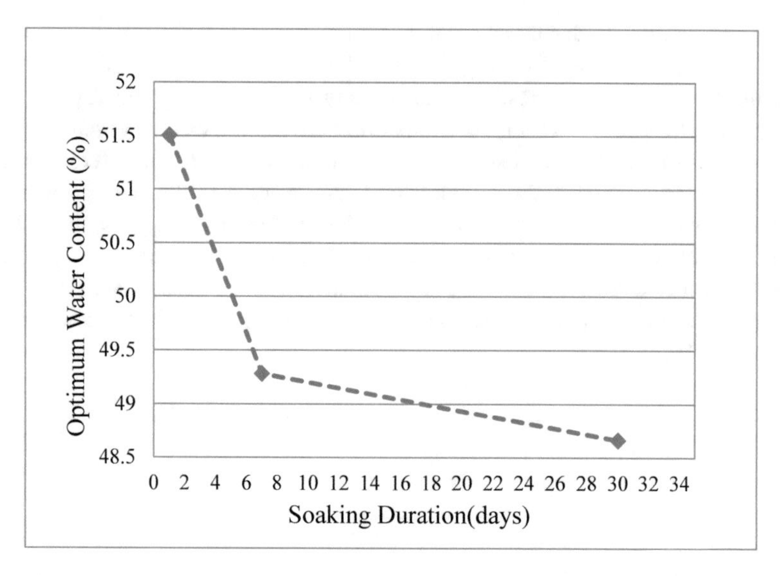

Fig. 10. Effect of the soaking duration on the optimum water content value

4.3 Effect of the Chemical Reaction Period on the Compaction Characteristics of the CKD

Figures 11 and 12 show the relation between both the maximum dry density and optimum water content, for the different allowed durations for chemical reaction of the CKD, after being mixed with a constant percentage of water content of 10%. The chemical reaction durations investigated are 1, 10 and 24 h. After the allowed periods for chemical reactions, the CKD samples were dried in the oven for 24 h under a temperature of 105 °C prior to performing the standard proctor compaction test. From Figs. 11 and 12, in addition to Table 4 summarizing the compaction tests results, it can be concluded that, in general, increasing the chemical reaction time decreases the resulting maximum dry density and increases the optimum water content. However, comparing with the case of performing the compaction test on the CKD samples directly on the dry CKD, without prior wetting with water and allowing certain durations for the chemical reaction, it can be seen that the wetting process resulted in decreasing the achieved maximum dry density and increasing the optimum water content. For the case of wetting with 10% w_c and allowing 1 h for the chemical reaction process, the maximum dry density decreased by about 4.78% (1.256 gm/cm^3 versus 1.319 gm/cm^3), while the optimum water content increased by about 21.08% (35.089% versus 28.979%), in comparison with the base case without prior wetting. For the case of wetting with 10% w_c and allowing 10 h for the chemical reaction process, the maximum dry density decreased by about 13.87% (1.136 gm/cm^3 versus 1.319 gm/cm^3), while the optimum water content increased by about 46.36% (42.415% versus 28.979%). In the third case of wetting with 10% w_c and allowing 24 h for the chemical reaction process, the maximum dry density decreased by about 15.69% (1.112 gm/cm^3 versus 1.319 gm/cm^3), while the optimum water content increased by about 53.96% (44.616% versus 28.979%).

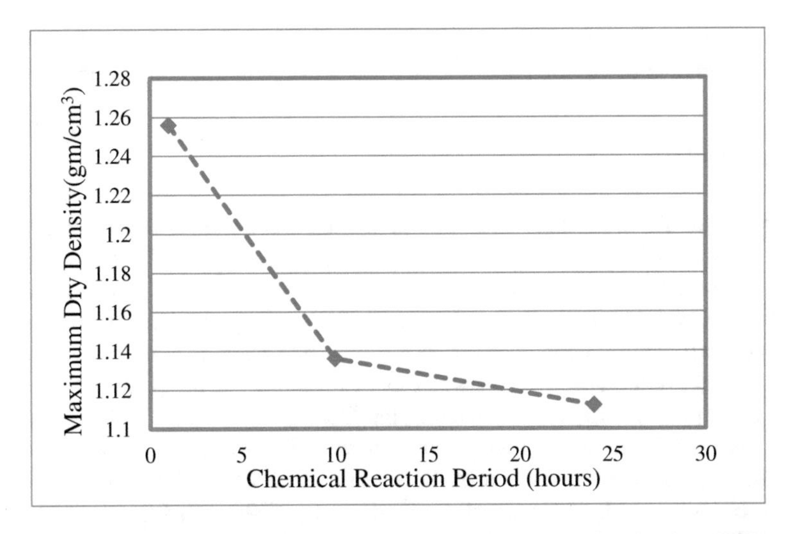

Fig. 11. Effect of the chemical reaction period on the Maximum dry density (CKD samples cured under 10% Wc prior to drying and compaction)

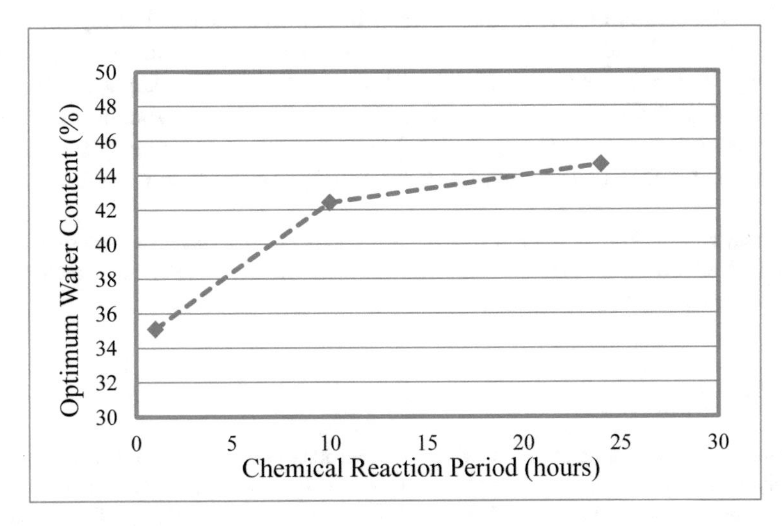

Fig. 12. Effect of the chemical reaction period on the optimum water content (CKD samples cured under 10% Wc prior to drying and compaction)

From these results, the maximum dry density is constantly decreasing by increasing the chemical reaction period for the CKD samples that are previously wetted with certain water content prior to drying and performing the compaction test. On the other hand, the optimum water content is generally increased by prior wetting and the degree of increasing is rising by increasing the time allowed for chemical reaction. Hence, it is not recommended to moist the CKD samples with water and allow any chemical reactions to take place prior to compacting the raw material in order to achieve the maximum possible dry density and the minimum possible optimum water content.

5 Conclusions

Using of waste oil as mixing liquid in compacting the CKD results in achieving the highest maximum dry density and lowest optimum fluid content. However, this technique is not recommended due to the anticipated high cost of the oil compared to either using the fresh water or the salt water/seawater. Also, using oil in the compaction of the CKD can entail environmental hazards that can be mitigated, however, using insulated landfills but at additional cost. Salt water as the seawater, if the plant is near to seas, provides an optimum way for compacting the CKD to the highest possible maximum dry density (1.342 gm/cm^3 for salt water versus 1.319 gm/cm^3 for fresh water) that means the least storage space at a reasonable optimum water content value that exceeds the case of using fresh water by about 4.64% only (30.323% for salt water versus 28.979% for fresh water).

Curing the CKD samples prior to compaction by either soaking (fully submergence) in water for certain periods or wetting and allowing the chemical reaction to take place for certain durations, didn't result in improving the efficiency of the compaction process of the CKD. In general, curing of the CKD with water resulted in decreasing the maximum dry density, which means the need for larger storage space, and increasing the optimum water content, requiring larger amount of water for compaction that is not economically feasible. Hence, there is no benefit of any prior curing of the CKD with water before performing the compaction process.

Acknowledgments. The used CKD material and corresponding chemical analysis results are provided by AL-ARABIA FOR CEMENT IN AIN SOKHNA, EGYPT. Their permission for using this material and provided information are gratefully acknowledged.

References

American Society of Testing and Materials, ASTM Standard Test Method for Chemical Analysis of Hydraulic Cement, Designation C 114 -03 ASTM International, West Conshohocken, PA

American Society of Testing and Materials, ASTM, Standard Test Method for Particle Size Analysis of Soils, Designation D 422 -02 ASTM International, West Conshohocken, PA

American Society of Testing and Materials, ASTM, (199). "standard test method for sieve analysis of fine and ciarse aggregates," C 136, Annual Book of ASTM Standards, ASTM International, West Conshohocken, PA

American Society of Testing and Materials, ASTM "Standard Test Method for SPECIFIC-GRAVITY-OF-SOLIDS DETERMINATION," Designation D 854-92 ASTM International, West Conshohocken, PA

American Association of State Highway and Transportation Officials, AASTO. Standard Method of Test for Moisture–Density Relations of Soils Using a 2.5-kg (5.5-lb) Rammer and a 305-mm (12-in.) Drop, Designation T-99

Baghdadi, Z.A., Fatani, N., Sabban, N.A.: Soil modification by cement kiln dust. J. Mater. Civ. Eng. ASCE **7**(4), 218–222 (1995)

Miller, G.A., Zaman, M., Rahman, J., Tan, K.N.: Laboratory and Field Evaluation of Soil Stabilization Using Cement Kiln Dust, Final Report, No. ORA 125- 5693, Planning and Research Division, Oklahoma Department of Transportation (2003)

Todres, H.A., Mishulovich, A., Ahmed, J.: Cement kiln dust management: permeability, Research and Development Bulletin RD103T, Portland Cement Association, Skokie, Illinois, USA (1992)

Author Index

S. Hemeda and M. Bouassida (Eds.): GeoMEast 2018, SCI, pp. 241–242, 2019.
https://doi.org/10.1007/978-3-030-01941-9

Printed in the United States
By Bookmasters